Sociology as a Life or Death Issue

FOURTH CANADIAN EDITION

ROBERT BRYM

University of Toronto

NELSON

Sociology as a Life or Death Issue, Fourth Canadian Edition

by Robert Brym

Publisher, Digital and Print Content:
Leanna MacLean

Marketing Manager:
Terry Fedorkiw

Content Manager:
Rachel Eagen

Photo and Permissions Researcher:
Jessie Coffey

Production Project Manager:
Jennifer Hare

Production Service:
MPS Limited

Copy Editor:
June Trusty

Proofreader:
MPS Limited

Indexer:
MPS Limited

Design Director:
Ken Phipps

Higher Ed Design PM:
Pamela Johnston

Cover Design:
Diane Robertson

Cover Image:
Aaron Millard

Compositor:
MPS Limited

Printed and bound in the United States
1 2 3 4 20 19 18 17

For more information contact Nelson Education Ltd., 1120 Birchmount Road, Toronto, Ontario, M1K 5G4. Or you can visit our Internet site at nelson.com

Library and Archives Canada Cataloguing in Publication Data

Brym, Robert J., 1951-, author
Sociology as a life or death issue / Robert Brym, University of Toronto. — Fourth Canadian edition.

Includes bibliographical references and index.
Issued in print and electronic formats.
ISBN 978-0-17-670004-1 (softcover).—ISBN 978-0-17-683088-5 (PDF)

1. Death—Social aspects—Textbooks. 2. Sociology—Textbooks. 3. Textbooks. I. Title.

HQ1073.B79 2017 306.9
C2017-901382-3
C2017-901383-1

ISBN-13: 978-0-17-670004-1
ISBN-10: 0-17-670004-8

In memory of my mother
Sophie Brym (1912–2004) ז"ל

and my father
Albert Brym (1911–2008) ז"ל

אַ ליכטיגע גן עדן זאָלן זיי האָבן

"This isn't fiction, Minny. It's sociology. It has to sound exact."
"But that don't mean it have to sound boring," Minny says.

–Kathryn Stockett, The Help *(2009: 418)*

"You know how everyone's always saying, 'Seize the moment'? I don't know, I'm kinda thinking it's the other way around. You know, like the moment seizes us."

–Nicole (Jessi Mechler), Boyhood *(2014)*

Table of Contents

FIGURES AND TABLES

Figures

Tables

The Fifth Dimension

We say an object is three-dimensional if it possesses length, width, and depth. It resides in the fourth dimension—time—if it exists for more than an instant. People usually experience the four dimensions effortlessly. You don't have to think much to know that you're holding a book in your hands and that time is passing as you read it, and most 14-year-old boys don't have to try hard to find the sleek shape of a shiny pair of Nike LeBron Soldier 10 basketball shoes pleasing and know that they are brand new.

In contrast, you need training to see the fifth dimension. Is it worth the effort? I invite you to read the next 160 or so pages and judge for yourself. I argue that the benefits are enormous. Seeing in five dimensions helps people live longer and fuller lives.

The fifth dimension is the *social*. Those new Nike LeBron Soldier 10s, which sell for $189.99, cost less than $10 to make in a crowded and poorly ventilated Indonesian factory. While LeBron James and major Nike shareholders get rich selling them, the 16-year-old girl who makes them works 15 hours a day for 50 cents an hour. Some college and university students find this situation deplorable and have organized an anti-Nike campaign and boycott to fight against it. Other college and university students note that the young factory worker probably prefers her job to the alternative—taking care of her siblings and the family goat back in her village for no pay and with no prospects at all—and that she is helping her poor country industrialize. From their point of view, buying Nike LeBron Soldier 10s benefits everyone.

Boycott or buy? Seen from the fifth, social dimension, an entire world of human relations and a moral dilemma are embedded in those basketball shoes. And what is true for Nike LeBron Soldier 10s is true for everything else in your life. Like it or not, you are part of society, and your actions, no matter how personal they appear, have consequences for others. It follows that if you want people to enjoy longer, richer lives, you need to make informed decisions about Nike LeBron Soldier 10s—and everything else you see before you.

That's where sociology comes in. **Sociology** is the systematic study of human action in social context. Sociologists analyze the social relations that lie beneath ordinary aspects of everyday life—everything from basketball shoes to homelessness, fame to racial discrimination, sex to religious zeal.

At its best, sociology speaks to the big issues of the day and the big issues of life, and it speaks to a broad audience. Accordingly, my aim in writing this little book was to speak plainly about the urgent need to think sociologically. Sociological understanding, I argue, is a life or death issue. I don't make that claim for dramatic effect. I mean it literally. By helping us understand the social causes of death, sociology can help us figure out how to live better; hence, the urgency of sociological knowledge.

I develop my argument in eight linked essays. In the title essay, I argue that it is useful to keep in mind the inevitability of death, because doing so compels us to focus on how to live best in our remaining time. I then outline how higher education in general and the sociological perspective in particular can contribute to that goal.

The following chapters add substance to the assertions of the opening essay. I examine death due to violence, disease, supposedly natural disasters, and racial and gender discrimination. I find that, in all of these cases, powerful social forces help to determine who lives and who dies. To make my case, I enter worlds that figure prominently in popular culture and the evening news—those of American hip hop, Palestinian suicide bombers, and victims of cancer and colonialism in Canada, hurricanes in the Caribbean region and the coast of the Gulf of Mexico, and of woman abuse, particularly in less developed countries in Africa and Asia.

Hip hop arose in desperate social circumstances that led to the death of many black men in particular. It was originally a protest against those circumstances. However, its message was diluted as the music became commercialized, and now it misguides many young people about how to best live their lives. A sociological understanding of hip hop holds out hope for correcting the problem.

Many Westerners think that suicide bombings are conducted by crazed religious zealots and that the appropriate response to them is overwhelming military force. My sociological analysis of Palestinian suicide bombings against Israelis shows that this way of thinking is flawed. The conflict between Palestinians and Israelis has deep political and emotional roots. Responding to suicide bombing with overwhelming military force only worsens the conflict and causes more death on both sides. Sociological analysis suggests an alternative approach.

When Hurricane Katrina killed about 2300 Americans in 2005, some people said the deaths were an awful consequence of a rare natural disaster, while others said they could have been avoided if the U.S. president hadn't been a racist. Sociologically speaking, these judgments are hopelessly naive. They do nothing to help us understand how disasters like Katrina can be prevented in the future. In contrast, my sociological analysis shows that patterns of class and racial inequality were responsible for most Katrina-related deaths and that other societies with different patterns of class and racial inequality have been more successful than the United States has been in avoiding such catastrophes. Again, sociology points the way to avoiding death and improving life.

Cancer is the leading cause of death in Canada. About 40 percent of women and 45 percent of men will develop cancer, and approximately 24 percent of women and 29 percent of men will die of the disease. People tend to think of cancer as a genetic problem, yet 90 percent or more of the genetic mutations that lead to cancer have environmental causes, including smoking, eating unhealthy food, and being exposed to toxic chemicals in the workplace and at home. In principle, we could remove all of these environmental causes and eliminate most cancers. Why we don't do that is a sociological puzzle that I investigate in Chapter 5. I find that class inequalities, industrial interests, and government inaction conspire to make cancer the scourge that it is, and I propose a series of policy innovations that could change this state of affairs.

The risks associated with being a woman differ over the life cycle and vary from one country to the next. In China and India, son preference is so strong that a high proportion of pregnant women use ultrasound to determine the sex of the fetus and undergo an abortion if it is female. In Pakistan and some other Muslim-majority countries, so-called "honour killings" are responsible for the death of thousands of women every year. In Canada, date rape is common and on the rise. Chapter 6 tries to explain variations in more than 136 countries in the level of danger that women face. It also highlights one of the most effective means of lowering "gender risk": passing laws that increase gender equality.

Europeans permanently settled in Canada in the early 1500s. Over the next four centuries, their actions (and the actions of subsequent immigrants) were directly responsible for the death of three-quarters of the Indigenous population. This is our national shame: Canadian genocide. Chapter 7 analyzes why European settlement and subsequent immigration had such disastrous consequences for Indigenous people, how the newcomers justified their actions, and what we need to do to rectify the many injustices committed in the name of progress.

Many students just beginning college or university want to know what sociology is and what they can do with a sociology degree. In the concluding essay, I draw on material from the preceding chapters to sketch the broad outlines of the discipline and offer career advice to undergraduates. I argue that my analyses of the social causes of death illustrate the fundamental aim of sociology at its best. Learning about the fifth dimension allows us to extend the third and fourth dimensions, helping us deepen and prolong life.

ACKNOWLEDGMENTS

Authors often say that they alone bear responsibility for their work. For two reasons, I make no such claim here.

First, I am a sociologist, and I therefore delight in acknowledging that I am embedded in intellectual and publishing networks whose members have helped to shape my work. I gladly share with them a full measure of responsibility for this book's strengths and weaknesses. Among the co-owners of this volume are colleagues who read and offered critical comments on all or part of the manuscript and provided useful bibliographic assistance: Bader Araj, Shyon Baumann, Maya Castle, Hae Yeon Choo, Jeff Denis, Adam Green, John Kirk, Anna Korteweg, Rhonda Lenton, Malcolm Mackinnon, Adie Nelson, Howard Ramos, Larissa Remennick, Lance Roberts, Jim Ron, Rania Salem, Hira Singh, Judy Taylor, Jack Veugelers, David Zitner, and nine anonymous reviewers selected by the publisher. I also want to single out my brilliant undergraduate research assistants, who helped me compose Chapter 5 and whose names are listed at the very beginning of Chapter 5. Culpable, too, are the members of the editorial, marketing, and production team at Nelson Education Ltd., who saw the value of my initial proposal and offered much encouragement and useful critical advice: Rachel Eagen, Leanna MacLean, Jennifer Hare, and June Trusty. I hope you all feel that the final product justifies your deeply appreciated efforts.

A journalist once asked the great Jewish poet Chaim Nachman Bialik (1873–1934) whether he preferred speaking Hebrew or Yiddish. Bialik answered in Yiddish: "Hebrew one speaks; Yiddish speaks by itself". I understand what he meant. Like all authors, I have struggled mightily on many occasions to get things right. This time, however, the job was almost effortless. That is the second reason I don't claim sole responsibility for this book: It practically wrote itself.

Robert Brym

Toronto

About the Author

Robert Brym (https://utoronto.academia.edu/RobertBrym) is SD Clark Chair of Sociology at the University of Toronto, a Fellow of the Royal Society of Canada, and the winner of numerous research and teaching awards, most recently the 2016 *British Journal of Sociology Prize*. He has published widely on politics and society in Russia, Canada, and the Middle East. He is currently researching social movement activism. Bob's popular introductory sociology textbooks have appeared in Canadian, American, Brazilian, Australian, and Quebec editions.

Courtesy of Bob Brym

Aaron Millard

1

Sociology as a Life or Death Issue

A DETOUR

To inspire you, I will take the unusual course of talking about death. I apologize in advance if this makes you uncomfortable. I know it is customary when addressing undergraduates to remind them that they are young, have accomplished much, and are now in a position to make important decisions that will shape the rest of their lives. I will eventually get around to that, too. But to arrive at the optimistic and uplifting part, I feel I must take a detour through the valley of the shadow of death.[1]

When I was seven years old, I lived across the street from a park where I engaged in all of the usual childhood games with my friends. We played tag, hide-and-seek, baseball, and cops-and-robbers. We also invented a game that we awkwardly called "See Who Drops Dead the Best." We would line ourselves up on a park bench and choose one boy to shoot the rest of us in turn, using a tree branch as a machine gun. Once shot, we did our best to scream, fall to the ground, writhe, convulse, and expire. The shooter would choose the most convincing

[1] This chapter is an expanded version of a commencement address delivered in May 2005 to graduates of the Steps to University program at the University of Toronto. Steps to University identifies promising senior high-school students who might otherwise not complete school or attend college or university because of their economic and social circumstances. It offers them selected university courses to encourage them to pursue postsecondary education.

victim—the boy who dropped dead the best—to play shooter in the next round. The game would occupy us for 10 minutes or so, after which we'd pick ourselves up and move on to baseball. At the age of seven, death was entertaining.

I didn't live in a war zone and there were no deaths in my family, so I really didn't begin to take death personally until I was 15. Then, one Sunday evening, it suddenly dawned on me that someday I would really die, losing consciousness forever. The moment this realization hit, I ran to my parents in panic. I rudely switched off the TV and asked them to tell me immediately why we were living if we were going to die anyway. My parents looked at each other, stunned, and then smiled nervously, perhaps thinking their son had taken leave of his senses. They were not especially religious people, and they had only a few years of elementary schooling between them. They had no idea how to address questions about the meaning of life. Eventually, my father confessed he didn't know the answer to my question, whereupon I ran to my bedroom, shouting that my parents were fools to have lived half a century without even knowing why they were alive. From that moment and for the next three decades, death became a source of anxiety for me.

DENIAL

And so it is for most adolescents and adults. We all know that we might die at any moment. This knowledge makes most of us anxious. Typically, we react to our anxiety by denying death. To a degree, denying death helps us calm ourselves.

The denial of death takes many forms. One is religious. Religion offers us immortality, the promise of better times to come, and the security of benevolent spirits who look over us. It provides meaning and purpose in a world that might otherwise seem cruel and senseless (James 1976 [1902]: 123, 139).

In one of its extreme forms, religion becomes what philosophers call **determinism**—the belief that everything happens the way it does because it was destined to happen in just that way. From the determinist's viewpoint, we can't really choose how to live because forces larger than us control life. Even religions that say we can choose between good and evil are somewhat deterministic because they guarantee eternal life only if we choose to do good, and that requires submitting to the will of God as defined by some authority, not by us. Many people worry less about death because they believe that the reward for submitting to the will of God is eternal life in heaven.[2]

A second way in which we calm our anxiety about death involves trying to stay young. Consider the cosmetic surgery craze. No official plastic surgery statistics are available for Canada, but in the United States in 2015, 15.9 million surgical and minimally invasive cosmetic procedures, including Botox injections, were performed. Since 2000, the overall number of procedures has risen 115 percent (American Society of Plastic Surgeons 2016).

And that's not all we do to stay young. We diet. We exercise. We take vitamin supplements. We wear makeup. We dye our hair. We strive for stylishness in our dress. We celebrate youthfulness and vitality in movies, music, and advertising. We even devalue the elderly and keep them segregated in nursing homes and hospitals, in part so we won't constantly be reminded of our own mortality.

The search for eternal youth is a form of what philosophers call **voluntarism**, the belief that we alone control our destiny. From the voluntarist's point of view, we can overcome forces larger than we are and thereby make whatever we want of our lives. Thus, many people worry less about death because they delude themselves into thinking they can cheat it.

A TRAP

I have good news and bad news for you, and I'm going to deliver the bad news first. The bad news is that the denial of death is a trap. Denying death makes it more difficult to figure out how to live well and thus be happy.

Let's say, for example, that a religion promises you eternal life in exchange for obeying certain rules. One rule says you can marry people only of your own religion. Another says that once you marry, you can't divorce. A third severely limits the steps you can take to control the number of children you have. A fourth says you have to marry someone of the opposite sex. Many people live comfortably guided by these rules, but the rules make others miserable. That, however, is the price they

[2] Secular versions of determinism also exist. Various forms of nationalism and communism promise a heavenly future for certain nations or classes. Paradoxically, however, they require that individuals submit to a higher party or state authority and act in prescribed ways if they hope to achieve what is supposedly historically inevitable (Berlin 2002a).

must pay for the religion's promise of eternal life. In general, by denying people the opportunity to figure out and do what is best for them as individuals, the deterministic denial of death can make some people deeply unhappy.

So can the voluntaristic denial of death. In *Nip/Tuck*, a popular TV series that ran from 2003 to 2010, plastic surgeons Christian Troy and Sean McNamara began each consultation with the words "Tell me what you don't like about yourself." Notice they didn't ask prospective patients what they disliked about their bodies. They asked them what they disliked about their selves. They assumed that the body faithfully represents the self—that your weight, proportions, colour, scars, and hairiness say something fundamentally important about your character, about who you are. If, however, we believe our happiness depends on our physical perfection and youthfulness, we are bound to be unhappy because nobody can be perfect and because we will inevitably grow old and die. In the meantime, pursuing youthfulness in the belief that you are no more than your appearance distracts you from probing deeply and finding out who you really are and what you need from life to make you happy. I conclude that denying death for whatever reason prevents you from figuring out how to live in the way that is best for you.[3]

HIGHER EDUCATION

Finally, some good news: You don't have to deny death, and thus become distracted from figuring out what you need to do to live a happy life. Instead, you can try to remain aware that you will die and that you could die at any moment. That awareness will inevitably cause you to focus on how best to achieve a meaningful life in your remaining time: the kind of career you need to pursue to make you happiest, the kind of person you need to develop a long-term intimate relationship with, the way you can best contribute to the welfare of others, the political principles you should follow, and so on. As an old saying goes, the gallows in the morning focuses the mind wonderfully (Frankl 1959).

I have more good news. People are well equipped to figure out how best to live. That is because we are meaning-creating machines.

[3] Some scientists believe that we will conquer death before this century is over by developing the ability to upload our minds to robots (Kurzweil 1999). If that happens, you may have plenty of time to revise my argument accordingly.

Faced with ambiguity in any social setting, we instantly start investing imaginative energy to define the situation and figure out what is expected of us and others. We abhor uncertainty, so we always strive to make social reality meaningful (Berger and Luckmann 1966). And since there is nothing more uncertain or ambiguous than death, when we face awareness of our own mortality, we almost instinctively want to create a durable purpose for our lives (Becker 1971, 1973). In fact, we are so eager to make life meaningful that we have created an institution especially devoted to helping us discover what the good life is for each of us: the system of higher education.

I imagine your parents and teachers have told you to stay in school as long as you can because a degree is a ticket to a good job. They are right, at least in part. A stack of studies shows that each additional year of education will increase your annual income for the rest of your life. People with more years of formal education tend to have jobs that give them more freedom and responsibility and allow them to be more creative, thus making work more enjoyable.

The view that colleges and universities are just places for job training is a half-truth, however. Above all, the system of higher education was developed as a place devoted to the discovery, by rational means, of truth, beauty, and the good life. Said differently, if you treat higher education not just as job training but as a voyage of self-discovery, you will increase your chance of finding out what you value in life, what you can achieve, and how you can achieve it.

Colleges and universities are divided into different departments, centres, schools, and faculties, each with a different approach to improving the welfare of humanity. The physician heals; the instructor in physical education teaches how to improve strength, stamina, and vigour; and the philosopher demonstrates the value of living an examined life. A good undergraduate education will expose you to many different approaches to improving your welfare and that of humanity as a whole, and will give you a chance to discover which of them suits you.

What does the sociological approach offer?

SOCIOLOGY

The sociological approach to improving human welfare is based on the idea that the relations we have with other people create opportunities for us not only to think and act but also set limits on our thoughts and actions. Accordingly, we can better understand what we are and what we can become by studying the social relations that help shape us.

A classic illustration of the sociological approach to understanding the world and improving human welfare is Émile Durkheim's late 19th-century study of suicide in France (Durkheim 1951 [1897]; Hamlin and Brym 2006). Most people think that suicide is the most non-social and anti-social action imaginable, a result of deep psychological distress that is typically committed in private and involves a rejection of society and everything it stands for. Yet Durkheim showed that high rates of psychological distress often do not result in a high suicide rate, while low rates of psychological distress sometimes do. He also argued that the **rate** and type of suicide that predominates in a society tells us something fundamentally important about the state of the society as a whole.[4]

According to Durkheim, the probability that your state of mind will lead you to suicide is influenced by the social relations in which you are embedded—in particular, the frequency with which you interact with others and the degree to which you share their beliefs, values, and moral standards. Durkheim referred to the frequency of interaction and the degree of sharing of beliefs, values, and morals in a group as its level of **social solidarity**.

In brief, Durkheim analyzed the effects of three levels of social solidarity on suicide rates (for more details, see the note accompanying Figure 1.1):

- *Low solidarity*. According to Durkheim, groups and societies characterized by a low level of social solidarity typically have a high suicide rate. Interacting infrequently and sharing few beliefs, values, and moral standards, people in low-solidarity settings lack emotional support and cultural guidelines for behaviour. They are therefore more prone to commit suicide if they experience distress.

  In his research, Durkheim found that married adults were half as likely as unmarried adults were to commit suicide because marriage typically created social ties and a moral cement that bound

[4] Dividing the number of times an event occurs (e.g., the number of suicides in a certain place and period) by the total number of people to whom the event could occur in principle (e.g., the number of people in that place and period), and then calculating how many times it would occur in a population of standard size (e.g., 100 000) will give you the rate at which an event occurs. Rates let you compare groups of different sizes. For instance, if 2 suicides occur in a town of 10 000 people and 4 suicides occur in a city of 100 000 people, the suicide rate is 20 per 100 000 in the town and 4 per 100 000 in the city.

Figure 1.1 Durkheim's Theory of Suicide

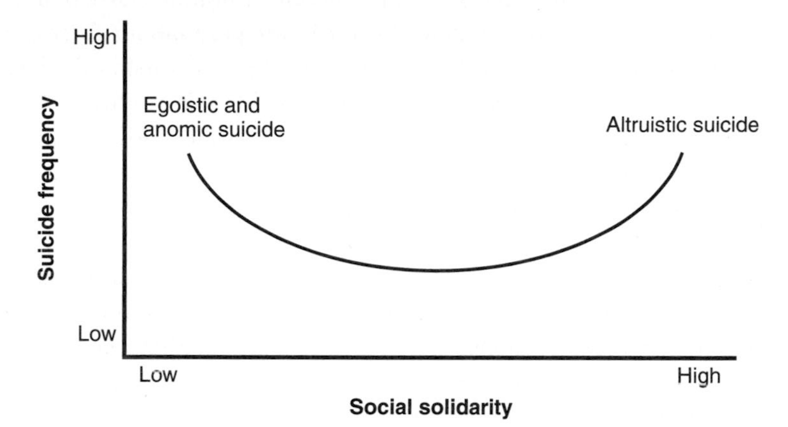

Note: Durkheim argued that, as the level of social solidarity increases, the suicide rate declines. Then, beyond a certain point, it starts to rise. Note the U-shaped curve in this graph. Durkheim called suicide that occurs in high-solidarity settings **altruistic**. In contrast, suicide that occurs in low-solidarity settings is egoistic or anomic. **Egoistic suicide** results from a lack of integration of the individual into society because of weak social ties to others. **Anomic suicide** occurs when norms governing behaviour become vaguely defined.

individuals to society. He found that women were only a third as likely as men were to commit suicide because they were generally more involved in the intimate social relations of family life. Jews were less likely to commit suicide than Christians were because centuries of persecution had turned them into a group that was more defensive and tightly knit. Elderly people were more prone than young and middle-aged people were to take their own lives when faced with misfortune because they were most likely to live alone, to have lost a spouse, and to lack a job and a wide network of friends.

On a broader, historical canvas, Durkheim viewed rising suicide rates as a symptom of the state of modern society. In general, social ties are weakening, he argued, and people share fewer beliefs, values, and moral standards than they used to.

- *Intermediate solidarity*. It follows that if we want suicide rates to decline, we must figure out ways of increasing the strength of social ties and shared culture in modern society. For example, if North Americans created a system of high-quality, universally accessible daycare, then more children would be better supervised, enjoy more interaction with peers and adults, and be exposed to similar socializing influences. At the same time, more adults (particularly single mothers) would be able to work in the paid labour force and form new social ties with their workmates. By thus raising the level of social solidarity, we would expect the suicide rate to drop.

- *High solidarity*. Despite a *general* decline in social solidarity in modern society, some groups are characterized by exceptionally high levels of social solidarity. When members of such a group perceive that the group is threatened, they are likely to be willing to sacrifice their lives to protect it. For instance, a soldier who is a member of a close-knit military unit may throw himself on a grenade that is about to explode to protect his buddies. Similarly, some suicide bombers see the existence of their group threatened by a foreign power occupying their homeland. They are willing to give up their lives to coerce the occupying power into leaving (Pape 2005). The increased rate of suicide bombing in the world since the early 1980s is, in part, a symptom of increasing threats posed to high-solidarity groups by foreign occupying forces. It follows that if we want fewer suicide bombings, one thing we can do is to figure out ways of ensuring that high-solidarity groups feel less threatened.

Much of the best sociological research today follows Durkheim's example. Sociologists frequently strive to identify (1) a type of behaviour that for personal, political, or intellectual reasons they regard as interesting or important, (2) the specifically social forces—the patterns of social relations among people—that influence that behaviour, and (3) the larger institutional, political, or other changes that might reliably improve human welfare with respect to the behaviour of interest. By conducting research that identifies these three elements, sociologists help people understand what they are and what they can become in particular social and historical contexts (Mills 1959).

WINNING THE GAME

You have accomplished much and are now in a position to make important decisions that will shape the rest of your life. At this threshold, I challenge you not to be seduced by popular ways of denying death. I challenge you to remain aware that life is short and that by pursuing a higher education, you will have the opportunity to figure out how to live in a way that will make you happiest. I personally hope you find sociology enlightening in this regard. But more important, you should know that higher education, in general, ought to encourage you to play the game of "See Who Lives Life the Best." You will be declared a winner if you play the game seriously. As Socrates once said to his pupils, "What we are engaged in here isn't a chance conversation but a dialogue about the way we ought to live our lives." Accept nothing less from your professors.

CRITICAL THINKING EXERCISES

1. Draw up a list of the three most important goals you want to achieve in life. Beside each goal, write a sentence about how you hope to achieve it. Exchange your list with a classmate. Write three paragraphs, each one addressing how worthwhile you think your classmate's goals is and whether the means he or she specified for achieving that goal is realistic. Discuss your evaluation with your classmate. Write two paragraphs that answer the following questions: What makes a life goal worthwhile? What makes a plan for achieving a life goal realistic?
2. List three sociological problems you would like to know more about. Write three paragraphs answering the following questions: What would you like to know about these sociological problems? Why are these problems important to you? How do you think you can find out more about them?
3. Durkheim discussed the association between the level of social solidarity and the likelihood of committing suicide. How do you think the level of social solidarity might influence the crime rate, the divorce rate, and a group's state of health? Why do you think social solidarity might have these effects? Answer in three paragraphs.

Aaron Millard

2

Hip Hop from Caps to Bling

STROLLING DOWN THE AVENUE

In January 2006, my wife and I attended a conference in New York. One afternoon we decided to take a break and walk over to Central Park. The weather was unusually pleasant for the time of year, and we enjoyed people-watching and window-shopping until we reached the pricey stores on Fifth Avenue. There, a barricade manned by two of New York's finest stopped our progress. About 10 m farther down the street, a second barricade blocked pedestrian traffic flowing in the opposite direction. The barricades cleared a space in front of Salvatore Ferragamo's flagship store, famous for its thousand-dollar shoes, stylish handbags, and other must-have accessories for the well-to-do.

Two black Lincoln Navigators were parked in front of the store. A high-tech garden of antennas and satellite dishes sprouted from the roof of one of them. Its tinted windows were shut. The windows of the second Navigator were wide open, and we could clearly see the driver and three other men inside, all clad in black. The passengers held AK-47 assault rifles upright. They wanted us to notice them. Two tall, athletic-looking men in suits, white shirts, ties, and well-tailored overcoats stood on either side of the store's front door, their eyes roaming the crowd. Each had his right hand inside his overcoat, presumably gripping a firearm. About half a dozen police vans and cruisers were blocking vehicular traffic. Police officers stood outside their cruisers.

Whatever emergency was in progress, an intimidating company of about two dozen well-armed men was positioned to deal with it.

"Hey!" I said to one of the men in blue. "Did you guys catch Osama bin Laden or something?"

The police officer suppressed a smile. "Not yet," he replied.

"So what's up?" I persisted.

"Can't say."

"Aw, come on. You can't stop all these taxpayers from enjoying the nice weather without an explanation. What's the occasion?"

"I guess Puffy needs a new tie."

"You mean Diddy, the hip hop artist?"

"Whatever."

In hip hop slang, "caps" are bullets and "bling" is flashy jewellery, as in "You don' hand over dat bling, I'ma bust a cap in yo' ass." Caps are the means and bling is the goal, as in the title of 50 Cent's movie, *Get Rich or Die Tryin'*. There was no slaughter on Fifth Avenue that fine January day, but I have to admit that, like the rest of the crowd, my wife and I were captivated by the staged threat of violence and the spectacle of material excess offered by Diddy's shopping excursion. We knew the show was contrived for publicity—Diddy is not the president of the United States and Fifth Avenue is not the inner city—but we were still excited to be close to the biggest revolution in youth culture since rock and roll, a revolution that unites death and wealth in a troubled marriage.

1.5 MILLION BLACK MEN ARE MISSING

Nobody should be surprised that a popular subculture rooted in the lives of African American men focuses so tightly on violence and death. More than a third of African American households enjoy annual incomes of US$50 000 a year or more. However, for the more than one-quarter of African Americans who live in poverty, violence and death are a big part of everyday life (U.S. Census Bureau 2016a).

One indicator of the disproportionate amount of violence faced by African American men is the **sex ratio**, the number of males per 100 females in a population. In the United States, there are 98 white American men for every 100 white American women (U.S. Census Bureau 2016b). This is the same sex ratio that one finds in Canada and approximately the same as in most other countries. There are two main reasons why there are more women than men in most countries. Men are more likely to work in dangerous jobs and engage in high-risk behaviour such as smoking and excessive alcohol consumption. Besides, women are the hardier sex, biologically speaking.[1]

Among African Americans, however, the sex ratio is just 91 (U.S. Census Bureau 2016a). Assuming that a sex ratio of 98 is approximately normal, we can conclude that 7 black men are "missing" for every 100 black women (since 98 minus 91 equals 7). Given the 21 million black women in the United States in 2004, that works out to about 1.5 million missing black men. In 2004, there were about 19 million black men in the United States, but there should have been around 20.5 million.

Many missing black men died violently. The **homicide rate** is the number of murders per 100 000 people in a population. The black male homicide rate was about 33 in 2012 but reached 49 in Indiana, 53 in Florida, 56 in Minnesota, and 59 in Montana. (In contrast, the homicide rate was 4.5 for the United States as a whole and 1.6 for Canada; Boyce and Cotter, 2013; Chokshi, 2015.) Figure 2.1 reinforces the point that homicide contributes to the low sex ratio among African Americans by plotting the homicide rate for black men against the black sex ratio for each state in the United States. It shows that, in states where relatively few black men are murdered, there are more black men than black women. But in states where many black men are murdered, there are fewer black men than black women (Boyce and Cotter 2013; Chokshi 2015).[2]

In 2014, 5720 more black men than black women were murdered in the United States. However, homicide is not the only cause of excess deaths among black men. In addition, 4966 more black men than black women died accidentally, mainly due to car accidents and drug overdoses. Some 1471 more black men than black women committed suicide and 1155 more black men than black women died of AIDS. Finally, 121 more black men than black women were killed by "legal intervention," that is, mainly by police shootings (calculated from National Center for Injury Prevention and Control 2016). These figures oblige us to conclude that the destruction of the lives of poor African American men by violence and high-risk behaviour is horrifyingly routine. It is, therefore, to be expected that violence and death would form central themes in their cultural expression (Centers for Disease Control and Prevention 2016b).

[1] In some countries, the sex ratio is unusually high; see Chapter 6 for an explanation.

[2] In Figure 2.1 and several other graphs in the book, I include a "trend line" that summarizes the relationship between the two variables in the graph. Technically, the trend line is known as the **least-squares regression line**. It is a straight line in a two-dimensional graph that is drawn so as to minimize the sum of the squared distances between each data point and the line itself.

Figure 2.1 Male Homicide Rate and Sex Ratio by State, African Americans, 2010–14

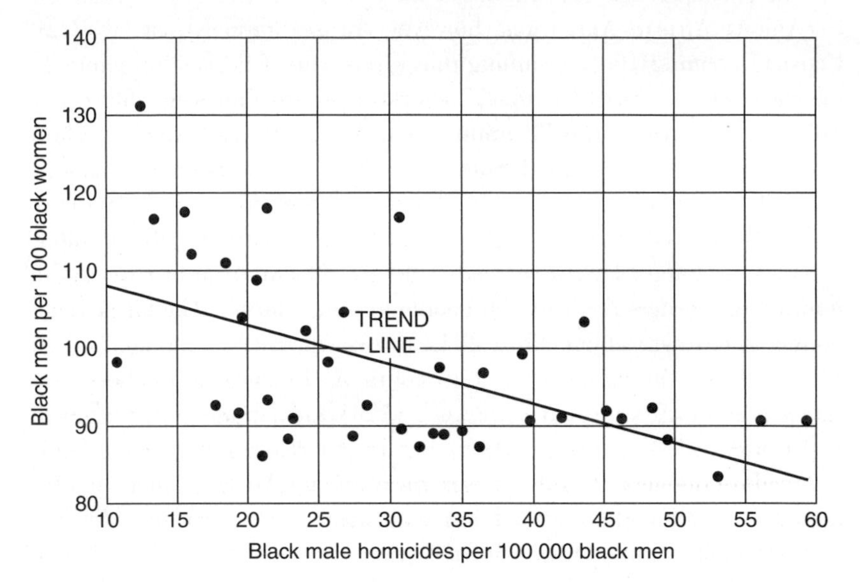

Note: Homicide data are not available for nine states, all with relatively small black populations. Sex ratios are for 2012.

Sources: Centers for Disease Control and Prevention. 2016b. "Fatal Injury Reports, 1999-2014, for National, Regional, and States (RESTRICTED). "http://webappa.cdc.gov/cgi-bin/broker.exe (retrieved 8 August 2016); US Bureau of the Census. 2016a. "Annual Estimates of the Resident Population by Sex, Race, and Hispanic Origin for the United States, States, and Counties: April 1, 2010 to July 1, 2015." http://factfinder.census.gov/faces/tableservices/jsf/pages/productview.xhtml?src=bkmk (retrieved 8 August 2016).

SOCIAL ORIGINS OF HIP HOP

The situation of the African American community as a whole has improved since the 1960s. The civil rights movement created new educational, housing, and job opportunities for African Americans and resulted in the creation of a substantial black middle class. The United States became a more tolerant and less discriminatory society. Yet in the midst of overall improvement, the situation of the more than one-quarter of African Americans who live in poverty became bleaker.

After World War II and especially in the 1960s, millions of southern blacks migrated to northern and western cities. Many were unable to find jobs. In some census tracts in Detroit, Chicago, Baltimore, and Los Angeles, black unemployment ranged from 26 to 41 percent in

1960. Many of the migrants were single women under the age of 25. Many had children but lacked a husband and a high-school diploma.

The race riots of the 1960s helped persuade the government to launch a "war on poverty," which increased the welfare rolls and worked for a time. In 1973, the American poverty rate fell to 11.1 percent, its lowest point ever. After 1973, however, everything went downhill. Manufacturing industries left the inner city for suburban or foreign locales, where land values were lower and labour was less expensive. In the three decades following 1973, the proportion of the American labour force employed in industry fell from about one-third to one-fifth. Unemployment among African American youth rose to more than 40 percent. Middle-class blacks left the inner city for the suburbs. The migration robbed the remaining young people of successful role models. It also eroded the taxing capacity of municipal governments, leading to a decline in public services. Meanwhile, the American public elected conservative governments at the state and federal levels. They cut school and welfare budgets, thus deepening the destitution of ghetto life (Piven and Cloward 1977: 264–361; Piven and Cloward 1993; Wilson 1987).

With few legitimate prospects for advancement, poor African Americans turned increasingly to crime and, in particular, to the drug trade. In the late 1970s, cocaine was expensive and demand for the drug was flat. So in the early 1980s, Colombia's Medellin drug cartel introduced a less expensive form of cocaine called "rock" or "crack." Crack was not only inexpensive, it also offered a quick and intense high and was extremely addictive. Crack cocaine offered many people a temporary escape from hopelessness and soon became wildly popular in the inner city. Turf wars spread as gangs tried to outgun each other for control of the local traffic. The sale and use of crack became so widespread that it corroded much of what was left of the inner-city African American community (Davis 1990).

The shocking conditions described above gave rise to a shocking musical form: hip hop. Stridently at odds with the values and tastes of both whites and middle-class African Americans, hip hop described and glorified the mean streets of the inner city while holding the police, the mass media, and other pillars of society in utter contempt. Furthermore, hip hop tried to offend middle-class sensibilities, black and white, by using highly offensive language.

In 1988, more than a decade after its first stirrings, hip hop reached its political high point with the release of the album *It Takes a Nation of Millions to Hold Us Back*, by Chuck D, frontman of Public Enemy.

In "Don't Believe the Hype," Chuck D (Carlton Douglas Ridenhour) accused the mass media of maliciously distributing lies. In "Black Steel in the Hour of Chaos," he charged the Federal Bureau of Investigation (FBI) and the Central Intelligence Agency (CIA) with assassinating the two great leaders of the African American community in the 1960s, Martin Luther King and Malcolm X. In "Party for Your Right to Fight," he blamed the federal government for organizing the fall of the Black Panthers, the radical black nationalist party of the 1960s. Here, it seemed, was an angry expression of subcultural revolt that could not be mollified.

HIP HOP TRANSFORMED

There were elements in hip hop that soon transformed it, however (Bayles 1994: 341–62; Neal 1999: 144–48). For one thing, early radical hip hop was not written as dance music. It, therefore, cut itself off from a large audience. Moreover, hip hop entered a self-destructive phase with the emergence of gangster rap, which extolled criminal lifestyles, denigrated women, and replaced politics with drugs, guns, and machismo. The release of Ice-T's "Cop Killer" in 1992 provoked strong political opposition from Republicans and Democrats, white church groups, and black middle-class associations. "Cop Killer" was not hip hop, but it fuelled a reaction against all anti-establishment music. Time Warner was forced to withdraw the song from circulation. The sense that hip hop had reached a dead end, or at least a turning point, grew in 1996, when rapper Tupac Shakur was murdered in the culmination of a feud between two hip hop record labels, Death Row in Los Angeles and Bad Boy in New York (Springhall 1998: 149–51).

If these events made it seem that hip hop was self-destructing, the police and insurance industries helped speed up its demise. In 1988, a group called Niggaz with Attitude released "Fuck tha Police," a critique of police violence against black youth. Law enforcement officials in several cities dared the group to perform the song in public, threatening to detain the performers or shut down their shows. Increasingly thereafter, ticket holders at hip hop concerts were searched for drugs and weapons, and security was tightened. Insurance companies, afraid of violence, substantially raised insurance rates for hip hop concerts, making them a financial risk. Soon, the number of venues willing to sponsor hip hop concerts dwindled.

While the developments noted above did much to mute the political force of hip hop, the seduction of big money did more. As early as 1982, with the release of Grandmaster Flash and the Furious Five's

"The Message," hip hop began to win acclaim from mainstream rock music critics. With the success of Run-DMC and Public Enemy in the late 1980s, it became clear there was a big audience for hip hop. It is significant that much of that audience was composed of white youths.

As one music critic wrote, they "relished ... the subversive 'otherness' that the music and its purveyors represented" (Neal 1999: 144). Sensing the opportunity for profit, major media corporations, such as Time Warner, Sony, Columbia, and BMG Entertainment, signed distribution deals with the small independent recording labels that had formerly been the exclusive distributors of hip hop CDs. In 1988, *Yo! MTV Raps* debuted. The program brought hip hop to middle America.

Most hip hop recording artists proved they were eager to forgo political relevancy for commerce. For instance, WU-Tang Clan started a line of clothing called WU Wear. With the help of major hip hop recording artists, companies as diverse as Tommy Hilfiger, Timberland, Starter, and Versace began to market clothing influenced by ghetto styles. Independent labels, such as Phat Farm and Fubu, also prospered. The members of Run-DMC once said that they "don't want nobody's name on my behind," but those days were long past. By the early 1990s, hip hop was no longer just a musical form but a commodity with spinoffs. Rebellion had been turned into mass consumption.

DIDDY

No rapper did a better job of turning rebellion into a commodity than Sean John Combs, better known as Puff Daddy, later as P. Diddy, and, as of 2005, Diddy. Diddy was born into a middle-class family in a New York suburb, became an avid Boy Scout, attended private school, played high-school football, and then enrolled in Howard University, the leading black university in the United States. Today, he is the world's richest hip hop artist, with a net worth estimated at US $735 million ("Top 100 Richest Rappers" 2016). Yet, despite his background, he seems to promote rebellion. For example, the liner notes for his 1999 CD, *Forever*, proclaim a revolution led by rebels, rule-breakers, and mavericks. Diddy says that by asserting your unique brand of individuality, you can change the world.

Note, however, that Diddy encourages only individual acts of rebellion, not radical, collective, political solutions. His politics became so mainstream that he became a prominent activist in the "get out the vote" campaign for the 2004 presidential election. Diddy's brand of dissent thus appeals to a broad audience, much of it white and middle class. As his video director, Martin Weitz, observed in an interview for *Elle* magazine, Diddy's market is not the inner city: "No ghetto kid from Harlem is going to buy Puffy. They think he sold out. It's more like the 16-year-old white girls in the Hamptons, baby!" (quoted in Everett-Green 1999).

It is also important to note that Diddy encourages individual acts of rebellion only to the degree that they enrich him and the media conglomerate he works for.[3] And rich he has become. Diddy lives in a multimillion-dollar mansion on Park Avenue in Manhattan and a multimillion-dollar house in the Hamptons. In 2005, *Forbes* magazine ranked him the 20th most important celebrity in the United States and the third biggest money earner among musicians, with an annual income of US$36 million ("The Celebrity 100" 2005). Diddy is entirely forthright about his self-enriching aims. In his 1997 song "I Got the Power," Diddy referred to himself as "that nigga with the gettin' money game plan" (Combs and the Lox 1997). And in *Forever*, he reminds us: "Nigga get money, that's simply the plan." From this point of view, Diddy has more in common with Martha Stewart than with Chuck D and Public Enemy (Everett-Green 1999).

BLING

Some hip hop artists, like Curtis "50 Cent" Jackson, come from the inner city, have criminal backgrounds, served time in prison, and glorify the gangster lifestyle. Some hip hop artists remain true to their political birthright. For example, Chuck D's mother was a Black Panther activist, and to this day he is engaged in raising black political consciousness as a writer, publisher, and producer. It seems, however, that a large number of prominent hip hop artists emulate the gangster lifestyle neither because it reflects their origins nor because they regard it as a political statement but simply because it is stylish and profitable to do so. Their backgrounds have nothing in common with drug suppliers, pimps, and gang leaders, and their politics are mainstream or non-existent. Even in the early years of hip hop, a gap between the biographies of many hip hop artists and their public personae was evident for those who took the time to do a background check. Three examples:

1. DMC (Darryl McDaniels) was part of the legendary Run-DMC, the first hip hop group that looked like it ran with a gang and

[3] *Forever* is marketed, manufactured, and distributed by a unit of BMG Entertainment, the multibillion-dollar entertainment division of Germany's Bertelsmann AG, the third-largest media company in the world.

had just come off the street corner. Run-DMC was credited with bringing hip hop into the mainstream in the 1980s. Yet DMC was born into a solidly middle-class, suburban family. His parents were college educated. He was described by rock critic Bill Adler as a good Catholic school kid, a mama's boy (Samuels 2004: 149).

2. Another infamous figure was Ice-T (Tracy Marrow). He is often credited with starting the gangster rap movement with his single, "6 'n the Morning." He released "Cop Killer" in 1992, causing a national scandal. Yet Ice-T completed high school and served in the army as a ranger in the 25th Infantry. He now continues serving the forces of good by playing a detective in the TV show *Law and Order: Special Victims Unit.*
3. Flavor Flav (William Johnathan Drayton, Jr.) was a member of notorious Public Enemy. Yet he graduated from high school and attended Adelphi, an old, respected college in Long Island, New York. He trained as a classical pianist. After a stint on the reality TV show *The Surreal Life*, he made a living co-starring in another reality TV show, *Strange Love*, with Brigitte Nielsen, a Danish actress once married to Sylvester Stallone. His next TV venture was *The Flavor of Love*, in which 20 single women who professed to adore him moved into a "phat crib" in Los Angeles and competed for his affections.

My contention that the "getting-money game plan" drives many hip hop artists is supported by the near-worship of luxury commodities in much of their music. From this point of view, hip hop is a lot like one of those soap commercials that rely mainly on brand-name repetition to ensure that consumers keep the product in mind when they go grocery shopping. Only in the world of hip hop, the good life is strongly associated not with laundry detergent but with driving a Mercedes, Bentley, Rolls-Royce, or Cadillac, drinking Hennessy cognac or Cristal champagne, and packing an AK-47 (Agenda Inc. 2005: 4–7).

STREET CRED

The runaway financial success of some hip hop artists can rob them of what they call "street cred." One's claim to be a pimp or a cop killer can lose credibility when one shops at Salvatore Ferragamo and lives in the suburbs.

Successful hip hop artists have responded to the problem of street cred in three ways. First, some decide to give up any pretence of street cred by using their money to insulate themselves from the inner city. A Diddy or a Will Smith (formerly the star of TV's *The Fresh Prince of Bel-Air*) makes no bones about catering to a largely white, suburban, culturally and politically mainstream, middle-class audience. They never lived in the inner city and apparently have no plans to visit anytime soon.

Successful hip hop artists whose audience appeal derives from their self-characterization as street toughs often take a more dangerous tack. They may live in the wealthy suburbs, but they still frequent the inner city, where some of them were born. Many of them are undoubtedly nostalgic about inner-city life, but they seem also to be motivated to visit the clubs and street corners of their old 'hood to show that they have not sold out. The trouble is that permanent residents often envy their wealth and fame, and this resentment can easily boil over into lethal violence. Famous hip hop artists who were shot and killed visiting their old neighbourhoods after striking it rich include Scott La Rock (Scott Sterling) in 1987 (the first high-profile hip hop slaying), Run-DMC's Jam Master Jay (Jason Mizell) in 2002, and Proof (DeShaun Holton), Eminem's right-hand man and member of D-12, in 2006 (Dawsey 2006). Wikipedia keeps a list of hip hop musicians who have been murdered since 1987. By far the leading cause of death: "Shot and killed" (Wikipedia 2016a).

A compromise between rejecting the inner city and visiting it as a rich tourist involves staging gun battles for public consumption. For example, in March 2005, a sidewalk gunfight broke out near hip hop radio station WQHT in New York City between the entourages of hip hop star The Game and his former mentor, 50 Cent. The Game had hinted that he might record with one of 50 Cent's rivals, so 50 Cent expelled The Game from his inner circle. The gunfight followed. Four years earlier, on the same street corner, a similar incident occurred between followers of Lil' Kim, one of the few female hip hop stars, and rival Capone after Capone's group had referred to Lil' Kim as "lame" in their appropriately titled song "Bang, Bang." In both gunfights, the hip hop stars' followers discharged many rounds of ammunition at close range, but damage was minor. Total casualties in the 2001 and 2005 gun battles combined: one man shot in the leg in 2005.

It seems plausible that these gunfights were for show. They helped reinforce the violent image and street cred of the hip hop stars involved. Hip hop stars are multimillionaire members of the music elite, but the gunfights confer "the illusion of their authenticity as desperate outlaws" (Hajdu 2005). In that light, shootouts are low-risk investments by savvy businesspeople. Lil' Kim's shootout certainly paid off handsomely. In front of a grand jury, she claimed that two of her associates were not present at the 2001 gunfight. But witnesses contradicted her testimony, and WQHT's security tape showed her holding a door open for one of the men. She was subsequently sentenced to a year and a day in prison for perjury. The two weeks preceding her imprisonment were videotaped for a reality show on Black Entertainment Television. The first episode of *Lil' Kim: Countdown to Lockdown* was the most watched series debut in the network's 25-year history. It has not been disclosed how much Lil' Kim earned for her efforts (Associated Press 2005; Strong 2006).

THE THREE PROMISES OF HIP HOP

Hip hop promises its followers three rewards: identity, upward mobility, and power. Let us consider each of these promises in turn.

Identity

> [M]an ain't like a dog ... because ... he know about death.... [W]e ain't gonna get no move on in this world, lyin' around in the sun, lickin' our ass all day ...,
> [S]o with this said, you tell me what it is you wanna do with your life.
>
> —*DJay (Terrence Howard), a pimp, to Nola (Taryn Manning), one of his prostitutes, in* Hustle and Flow *(Film, 2005)*

People create, share, and socially transmit languages, beliefs, symbols, values, material objects, routine practices, and art forms to help them survive and prosper. Sociologists call the sum total of these responses to real-life problems **culture**. Medicine, Christianity, the Russian language, and the pulley help people cope, respectively, with ill health, questions about the meaning of life, the desire to communicate, and the need to raise heavy objects. Hip hop is no different. It is also a response to real-life problems (Swidler 1986).

For example, the 2005 box-office hit *Hustle and Flow* tells the story of how DJay responds culturally to his life problems. Knowing that we will die, we must choose how to live meaningfully or be reduced to an existence little better than that of a dog, says DJay. He finds that he can achieve self-fulfillment by giving up his life as a pimp and giving voice to the joys and frustrations of the life he knows in the largely black, poor, violent, downtown core of Memphis, Tennessee. He becomes a hip hop artist. Artistic self-expression renders his life meaningful and rewarding. It gives him a sense of identity.

To operate in the world, all people must develop a sense of who they are and what they can do (and who they aren't and what they can't do). The construction of identity is a lifelong task; people may alter their occupational, religious, national, ethnic, and even sexual identity as they mature and their circumstances change. But adolescence is the stage of life when most people lay the foundation for future development. It is typically a turbulent period, full of tentative experiments, exuberant strivings, the emulation of heroes, self-doubt, false starts, and confrontation with stubborn authority. By means of these experiments, strivings, and so on, adolescents form a baseline identity. Particular styles of popular music—unique patterns of rhythm, melody, and lyrics—express adolescent struggles in particular social contexts and give them form. That is why popular music is so meaningful and important to most adolescents (and nostalgic adults) (Gracyk 2001).

Some current hip hop songs oppose violence, crime, drugs, and the mistreatment of women, but the dominant identity promoted by the genre is that of proud, arrogant, violent, criminal, misogynistic, black hyper-masculinity (Weitzer and Kubrin 2009). The identity is largely a response to the degrading effects of racism on the self-esteem of black men in the American inner city. Take persistent poverty and bad schools, remove social services and industrial jobs, introduce crack and gang wars, and you soon get hip hop (Dyson 2004). Nelson George, the genre's leading historian, writes that hip hop is "a system of survival" and "an invigorating source of self-empowerment" (George 1999: 50). It negates middle-class sensibilities because many black men believe that middle-class sensibilities have tried to negate them.

Upward Mobility

George is correct to note that "hip hop didn't start as a career move but as a way of announcing one's existence in the world" (George 1999: 14). Nonetheless, a career move it soon became. If hip hop's first promise was to provide a sense of black male identity in the context of the American inner city in the 1970s and 1980s, its second promise was to serve as a path of upward social mobility out of that context. (**Upward mobility** refers to movement up a system of inequality.)

Yet hip hop's lure resembles the largely false hope offered by professional basketball and football (see Table 2.1). Even taking soccer, hockey, and baseball into account, North American professional sports employed 5267 players in 2013–15, 2418 of whom were black. That's 2418 out of more than 6 million black men between the ages of 20 and 39. The odds of an African American man in the 20–39 age cohort being a

Table 2.1 African American Men in Professional Sports, 2013–16

Sport	Players	Black Players	Blacks as Percentage of Total
National Football League	2841	1883	67.3
National Basketball Association	446	332	74.4
National Hockey League	690	80	11.6
Major League Soccer	540	61	11.1
Major League Baseball	750	62	8.3
Total	5267	2418	45.9

Sources: Yahoo! Answers. "How many players the NHL has currently?"; Richard Lapchick, 2016; Wikipedia, 2016b, "List of black NHL players."

top professional athlete are about 2500 to 1. If he lives to the age of 80, he has about the same chance (3000 to 1) of getting struck by lightning in his lifetime (estimated from "Facts about Lightning" 2006; U.S. Census Bureau 2002). Although I have not been able to find statistics on the subject, it is evident that the odds of an African American man becoming a hip hop star are considerably worse than his odds of becoming a top professional athlete; the black men who become well-known hip hop artists even at the regional level, let alone nationally or internationally, number in the low hundreds, not the low thousands.

The poor black youth who regard professional athletes and hip hop artists as role and mobility models have little chance of realizing their dreams, all the more so because their unrealistic aspirations often deflect their attention from a much safer bet: staying in school, studying hard, and pursuing an ordinary career (Doberman 1997). The odds of an African American man in the 18–40 age cohort being a physician are roughly seven times better than the odds of his being a professional athlete or a well-known hip hop artist, and the odds of his being a lawyer are roughly 14 times better (estimated from Holmes 2005; King and Bendel 1995; U.S. Census Bureau 2002). Yet because so many young African American men seek to follow the career paths and emulate the lifestyles of a 50 Cent or a LeBron James, too few of them sing the praises of, say, Dr. James McCune Smith, the first African American doctor. In 2005, the number of black law students in the United States fell to a 12-year low despite a growing black population (Holmes 2005).

An important lesson about the nature of culture lies embedded in this story. Culture is created to solve human problems, as we have seen. But not all elements of culture solve problems equally well. Some elements of culture even create new problems. After all, the creators of culture are only human. In the case at hand, it seems that by promoting unrealistic hopes for upward mobility and encouraging a lifestyle that draws young African American men away from school, hard work, and the pursuit of an ordinary career, hip hop culture badly short-changes them.

Power

Like hip hop's promise of upward mobility, its assurance of power has proven largely an illusion.

We saw that hip hop emerged among African American inner-city youth as a counsel of despair with strong political overtones. Many commentators believed that by reflecting the traditions, frustrations, and ambitions of the community that created it, hip hop would help the otherwise isolated voices of poor black youth sing in unison, shape a collective identity, and engage in concerted political action to improve the conditions of all African Americans (cf. Mattern 1998).

There are still radical currents in hip hop. For example, in 2012, Macklemore and Lewis's first hit, "Same Love," criticized homophobia in hip hop and promoted gay rights. Their single, "Thrift Shop," which topped the Billboard "Hot 100" for six weeks and won a Grammy Award for best rap song in 2014, is a critique of mindless consumerism. Nonetheless, such currents seem to be minor, at least in North America.

We can reasonably assume that, among hip hop artists, popularity and wealth are strongly correlated. Examining a list of the 100 wealthiest hip hop artists, whose net worth ranges from US$10 million to US$750 million, yields the following findings ("Top 100 Richest Rappers 2016). With the exception of Canada's Drake (ranked 17th), all 100 are from the United States. Only five are women, with Nicki Minaj occupying the top women's position at number 23. Only one may be classified as politically radical or socially conscious—Nas, who ranks 60th. Other socially conscious hip hop artists, such as Macklemore and Lewis, Immortal Technique, Mos Def, and Talib Kweli are not on the list.

Radicalism in hip hop seems to be more common outside North America. In Senegal, the playing of hip hop that was highly critical of the government was widely believed to have helped topple the ruling party in the 2000 election. In France, North African youth living in impoverished and segregated slums use hip hop to express their political discontent, and some analysts say the genre helped mobilize youth for anti-government rioting in 2005 (Akwagyiram 2009). "Rais Lebled" [Mr. President], a song by Tunisian rapper El Général, became the anthem of young people participating in the democratic uprisings in Tunisia, Egypt, and elsewhere in the Arab world in 2011: "Mr. President, your people are dying/People are eating rubbish/Look at what is happening/Miseries everywhere, Mr. President/I talk with no fear/Although I know I will get only trouble/I see injustice everywhere" (Ghosh, 2011).

However, in North America, hip hop has become, for the most part, an apolitical commodity that increasingly appeals to a racially heterogeneous, middle-class audience. As one of hip hop's leading analysts and academic sympathizers writes, "the discourse of ghetto reality or 'hood authenticity remains largely devoid of political insight or progressive intent" (Forman 2001: 121).

Hip hop substantially lost its politics for three reasons. First, as one industry insider notes, "Mainstream media outlets and executive decision-makers ... fail to encourage or support overt political content and militant ideologies because ... 'it upsets the public'" (KRS-One, cited in Forman 2001: 122). The recording industry got excited about hip hop precisely when executives saw the possibility of "crossover," that is, selling the new black genre in the much larger white community. For them, hip hop was an opportunity little different from that offered by Motown in the 1960s. They apparently understood well, however, that to turn hip hop into an appealing mass-marketed commodity, it had to be tamed and declawed of its political content so as not to offend its large potential audience. If they needed to be sensitized to the need to tone down the rhetoric, the political opposition to hip hop that was stimulated by gangster rap and songs like "Cop Killer" in the early 1990s certainly helped. That opposition was the second reason hip hop lost its politics.

Third, hip hop artists themselves contributed to the depoliticization of their music. For the most part untutored in politics, history, and sociology, they are not equipped to think clearly and deeply about the public policies that are needed to help the black underclass and the forms of political action that are needed to get the black underclass to help itself. At most, they offer the flavour of rebelliousness, the illusion of dissent, giving members of their audience the feeling of being daring and notorious rule-breakers and revolutionaries but offering nothing in the way of concrete ideas, let alone leadership.

Vladimir Lenin, leader of the Russian Revolution of 1917, once said that capitalists are so eager to earn profits, they will sell the rope from which they themselves will hang. But he underestimated his opponents. Savvy executives and willing recording artists have taken the edge off hip hop to make it more appealing to a mass market, thus turning dissent into a commodity (Frank and Weiland 1997). Young consumers are fooled into thinking they are buying rope to hang owners of big business, political authorities, and cultural conservatives. Really, they're just buying rope to constrain themselves.

CULTURE AND SOCIAL STRUCTURE

Social structures are relatively stable patterns of social relations that constrain and create opportunities for thought and action. For example, in a social structure composed of just two people, both individuals must be engaged for the structure to persist. If one person fails to participate and contribute to the satisfaction of the other, the relationship will soon dissolve. In contrast, three-person social structures are generally more stable because one person may mediate conflict between the other two. In addition, three-person structures allow one person to exploit rivalry between the other two in order to achieve dominance. Thus, the introduction of a third person makes possible a new set of social dynamics that are impossible in a two-person structure (Simmel 1950). What is true for two- and three-person relationships holds for social structures composed of millions of people who are organized into institutions, racial groups, social classes, and entire societies; they constrain and create a host of opportunities for the people who comprise them.

Sociologists have long debated whether social structure gives rise to culture or vice versa. At first glance, this may seem to be a chicken-and-egg issue. For instance, in this chapter, I have argued that the social structure of the American inner city gave rise to the cultural phenomenon of hip hop. On the other hand, I have also argued that the culture of hip hop, insofar as it encourages violence and diverts attention from more realistic avenues of social mobility, reinforces the social structure of the American inner city. Chicken or egg?

Research on Media Violence

Since the 1960s, social scientists have employed the full range of sociological methods to investigate the effects of the mass media on real-world behaviour. Most of their research focuses on *violent* behaviour, and it's worth reviewing because it can improve our understanding of the relationship between hip hop culture and inner-city social structure.

Some of the research is based on **experiments**, carefully controlled artificial situations that allow researchers to isolate presumed causes and measure their effects precisely. In a typical experiment, a group of children is randomly divided into "experimental" and "control" groups. The experimental group is shown a violent TV program. The level of aggressiveness of both groups at play is measured before and after the showing. If, after the showing, members of the experimental group play significantly more aggressively than they did before the showing, and significantly more aggressively than members of the control group, the researchers conclude that TV violence affects real-world behaviour.

Scores of such experiments show that exposure to media violence increases violent behaviour in young children, especially boys, over the short term. Results are mixed when it comes to assessing longer-term effects, especially on older children and teenagers (Anderson and Bushman 2002; Browne and Hamilton-Giachritsis 2005; Freedman 2002).

Sociologists have also used surveys to measure the effect of media violence on behaviour. In a **survey**, randomly selected people are asked questions about their knowledge, attitudes, or behaviour. Researchers aim to study part of a group (a **sample**) to learn about the whole group of interest (the **population**). The results of most surveys show a significant relationship between exposure to violent mass media and violent behaviour, albeit a weaker relationship than experiments show. In a survey of Toronto high-school students who listened almost exclusively to hip hop, researchers found that involvement in property and violent crime was strongly associated with appreciation of hip hop among white and Asian youth but not among black youth (Tanner, Asbridge, and Wortley 2009: 709). Similar surveys, however, find no such relationship (Anderson and Bushman 2002; Huesmann, Moise-Titus, Podolski, and Eron 2003; Johnson et al. 2002).

Field research—systematically observing people in their natural social settings—has also been employed to help us understand how media violence may influence behaviour. For example, sociologists have spent time in schools where shooting rampages have taken place. They have developed a deep appreciation of the context of school shootings by living in the neighbourhoods where they occur; interviewing students, teachers, neighbourhood residents, and shooters' family members; and studying police and psychological reports, the shooters' own writings, and other relevant materials (Harding, Fox, and Mehta 2002; Sullivan 2002). They have tentatively concluded that only a small number of young people—those who are weakly connected to family, school, community, and peers—are susceptible to translating media violence into violent behaviour. Lack of social support allows their personal problems to become greatly magnified, and if guns are readily available, they are prone to using violent media messages as models for their own behaviour. In contrast, for the overwhelming majority of young people, violence in the mass media is just a source of entertainment and a fantasy outlet for emotional issues, not a template for action (Anderson 2003).

Finally, **official statistics** (numerical data originally compiled by state organizations for purposes other than sociological research) have been analyzed to place the effect of media violence on real-world behaviour into broader perspective. For example, researchers have discovered that between 1989 and 2000, the percentage of popular hip hop songs mentioning homicide in their lyrics increased from 29 percent to 42 percent. However, the homicide rate (an official statistic) fell substantially during this period. This result is just the opposite of what one would expect if hip hop had a positive influence on the homicide rate (Hunnicutt and Andrews 2009: 619).

The big differences in violent behaviour between the United States and Canada are also relevant. The homicide rate has historically been about three to four times higher in the United States. Yet TV programming, movies, and video games are nearly identical in the two countries, so exposure to media violence can't account for the difference. Most researchers attribute the difference to the higher level of economic and social inequality and the wider availability of handguns in the United States (Government of Canada 2002; Lenton 1989; National Rifle Association 2005).

My literature review leads me to unscramble this particular chicken-and-egg debate as follows. Media violence in general, and hip hop culture in particular, probably stimulate real-world violence among a minority of young people, although to a considerably lesser degree than some alarmists would have us believe (McWhorter 2005: 315–51). The effects are strongest among male adolescents who lack strong ties to family and other institutions that, by example, instruction, and discipline, typically socialize young people to refrain from violence.

They are especially exaggerated in settings where economic and social inequality is high and where handguns are readily available.

However, the effects of the mass media on *non-violent* behaviour are probably more widespread and stronger still. After all, parents, teachers, and religious figures spend a lot of time and effort teaching children and adolescents that violence is, in most cases, wrong. Along with the police and the courts, they impose severe penalties on youth who act violently. In contrast, while parents and teachers may look askance at a pair of pants that hang far below the waist, few people are likely to impose penalties for wearing them. That's why the mass media probably exert more influence on lifestyle than on violence.

I conclude that hip hop culture probably does a more effective job getting young people to dress and talk in certain ways and misguiding them about their mobility prospects than it does influencing them to act violently. It follows that social reformers interested in lowering levels of violence can achieve little by bashing hip hop culture. They could lower the level of violence a lot more effectively by figuring out ways to limit the availability of handguns and shore up or provide alternatives to faltering social institutions in the inner city, especially schools and families.

CRITICAL THINKING EXERCISES

1. What do you personally gain from hip hop? How are you negatively affected by hip hop? Do the positive influences outweigh the negative influences or vice versa? After drawing up a balance sheet for yourself, create a similar balance sheet for society as a whole. How, if at all, does your personal balance sheet differ from your societal balance sheet?
2. In a page, explain why, in your opinion, different research methods reach somewhat different conclusions about the effects of media violence on real-world behaviour. How can the research method itself influence the results of the research?
3. Some analysts argue that social conditions are wholly responsible for the behaviour of people. Others argue that people are responsible for their own behaviour. Still others hold that while social conditions influence behaviour, people still have plenty of room for choice and therefore bear some responsibility for their actions. In two pages, justify your position. Make your case based on the example of African Americans or Aboriginal Canadians.

Aaron Millard

3

Explaining Suicide Bombers

FROM KARBALA TO MEKHOLA

"Ya Karbala! Ya Hussein! Ya Khomeini!" That was the cry of waves of Iranian children and youths armed with Kalashnikovs and hand grenades as they attacked Iraqi positions during the early years of the Iran–Iraq war (1980–88). Entrenched machine guns and helicopter gunships mowed them down, but new waves kept on coming (Baer 2006; Reuter 2004). They fought for Karbala, the town where, in 680 CE,[1] a battle took place between factions that disagreed over how they should choose the successor to the Prophet Muhammad; the Sunni wanted the successor to be elected from a certain tribe, while the Shi'ites wanted him to be Muhammad's direct descendant. They fought in the name of Hussein, Muhammad's grandson, who led vastly outnumbered forces in a suicidal battle against the attacking Sunni at Karbala. Finally, they fought for Khomeini, the spiritual and political leader of Shi'ite Iran in the early 1980s. Lacking weapons and a well-organized army, Khomeini proclaimed it an honour to die in holy battle and instructed recruiters to find human cannon fodder

Parts of this chapter are based on Brym (2007, 2008) and Brym and Araj (2006, 2008).

[1] "CE" stands for "common era" and is now preferred over the ethnocentric AD, which stands for *anno Domini* (Latin for "the year of our Lord").

in Iran's schools. In this way, suicide attacks were institutionalized as a technique of collective violence in the modern Islamic world.

Suicide attacks were moulded into a precision instrument shortly afterward, when about 2000 Iranian Revolutionary Guard militants arrived in southern Lebanon to support the anti-Western and anti-Israel Hizballah movement (see Figure 3.1). The first suicide bombing against Western interests in the Middle East took place in Beirut, Lebanon, in October 1983, when Shi'ite militants attacked the military barracks of American and French peacekeepers, killing nearly 300 people. Four months later, Western troops fled the country, thereby teaching the attackers that suicide bombings could not only be inexpensively organized, accurately directed, and precisely controlled, but that under some circumstances, they could yield quick and substantial payoffs.

Israel had invaded Lebanon in the summer of 1982 in an effort to crush the Palestine Liberation Organization (PLO), which sought to recapture territory won by Israel in earlier wars (Brym 1983; see Figure 3.2). The Israelis succeeded in forcing the PLO leadership out of Lebanon. However, the Iranian-backed Hizballah and several copycat groups used the presence of Israeli troops in Lebanon as an opportunity to launch more suicide attacks.

In 1985, Israel partially withdrew from Lebanon. It now sought to weaken the PLO in the West Bank and the Gaza Strip, territories it had occupied since its 1967 war with its Arab neighbours (see Figure 3.2). To that end, Israel permitted the establishment of a conservative Islamic organization—the Islamic Resistance Movement, or Hamas. It judged that the new organization would serve as a moderate political counterweight to the PLO. Israel let Hamas accept funding from Saudi Arabia, turned a blind eye as its supporters stormed cinemas and set fire to restaurants selling alcohol, and allowed the creation of the Islamic University of Gaza. Ironically, the university later became a recruiting ground for suicide bombers (Reuter 2004: 98).

Hamas became the leading proponent of suicide bombings inside Israel and its occupied territories. In April 1993, the first such attack took place in the rural Israeli settlement of Mekhola. Nineteen similar attacks were staged over the next four years in Israel, the West Bank, and Gaza. Between 1993 and 1997, suicide bombers were responsible for the death of 175 people (including 21 suicide bombers) and the injury of 928 others (Johnston 2003). A second and more lethal wave

Figure 3.1 The Middle East

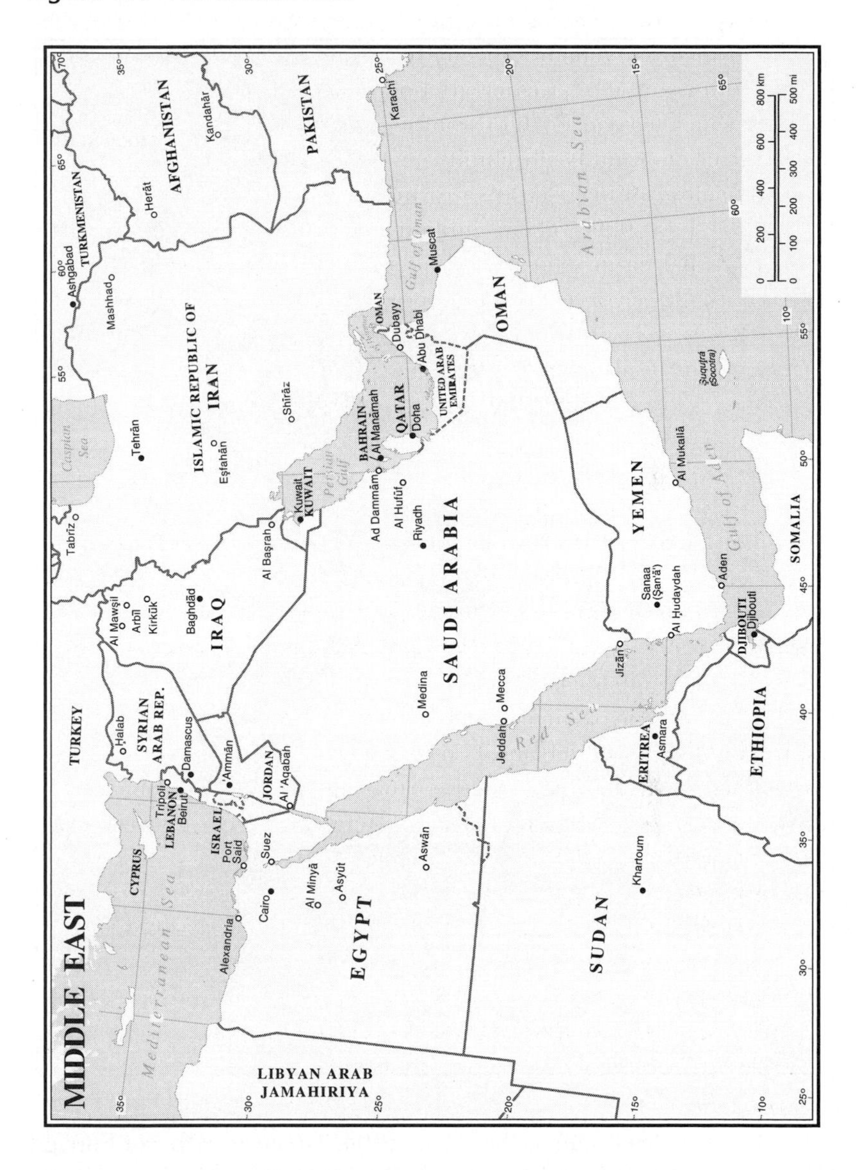

Figure 3.2 Israel, the West Bank, and Gaza

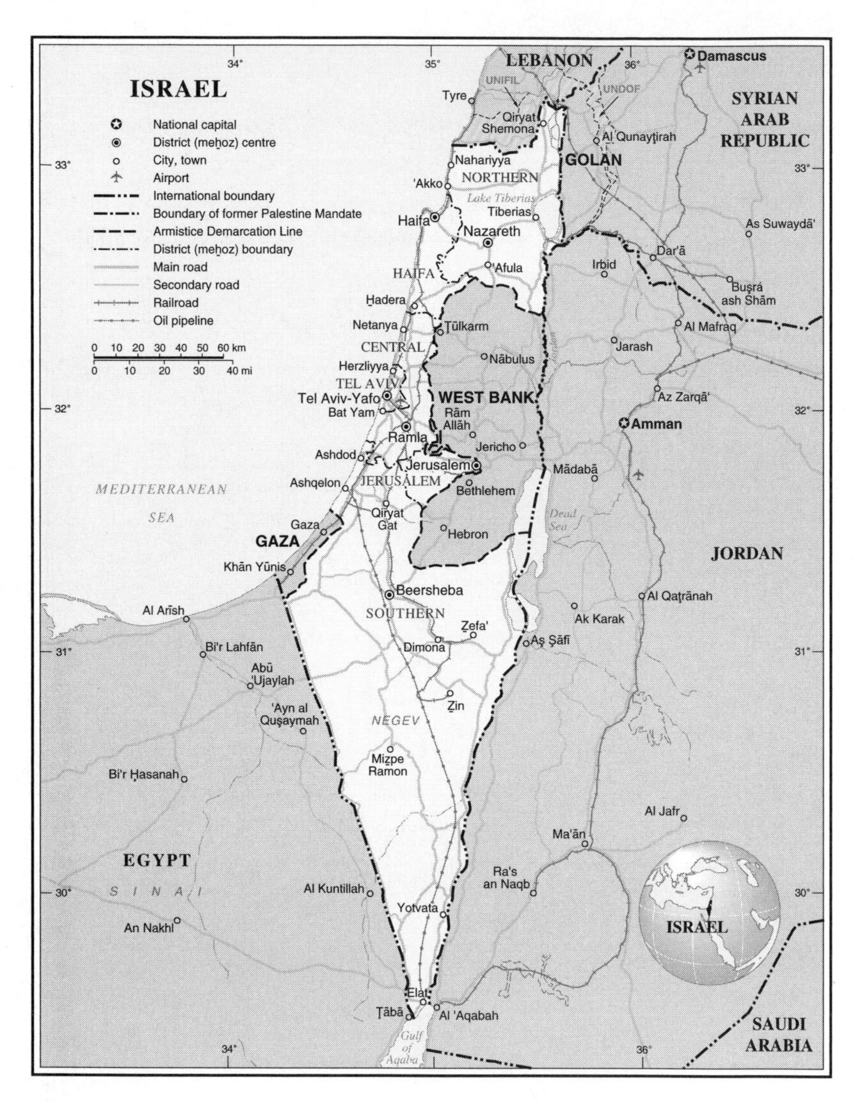

Policy Institute for Counter-Terrorism 2004; Israeli Ministry of Foreign Affairs 2004; *New York Times* 2000–2006). In the early years of the 21st century, Israel, the West Bank, and Gaza became the region of the world with the highest frequency of, and the highest per capita death toll due to, suicide bombing.

EXPLANATIONS

Explanations for the rise in suicide bombing since the early 1980s focus on three sets of factors:

1. the characteristics of the suicide bombers
2. the occupation, by perceived foreigners, of territory claimed by the suicide bombers and their organizations as their homeland
3. the interaction between perpetrators of suicide attacks and occupiers

Let us examine each of these explanations in turn.

Focus on the Perpetrators

Psychopathology

Lance Corporal Eddie DiFranco was the only survivor of the 1983 suicide attack on the U.S. Marine barracks in Beirut who saw the face of the bomber. DiFranco was on watch when he noticed the attacker speeding his truck full of explosives toward the main building on the marine base. "He looked right at me [and] smiled," DiFranco recalls (quoted in Reuter 2004: 53).

Many Western observers quickly passed a verdict: People willing to blow themselves up to kill others must be abnormal, and if they die happily, they must surely be deranged. Several psychologists characterized the Beirut bombers as "unstable individuals with a death wish" although they lacked any evidence of the bombers' state of mind (Perina 2002). Similarly, following the September 11, 2001, suicide attacks on the United States, the U.S. government and media interpretations underscored the supposed irrationality and even outright insanity of the bombers, again without the benefit of supporting data (Atran 2003: 1535–36).

Despite such claims, interviews with prospective suicide bombers and reconstructions of the biographies of successful suicide bombers do not suggest a higher rate of psychopathology than in the general population (Brym and Araj 2012a, 2012b; Davis 2003; Reuter 2004; Stern 2003; Victor 2003). A study of all 462 suicide bombers who attacked targets worldwide between 1980 and 2003 found not a single case of depression, psychosis, past suicide attempts, and so forth, among them, and only one case of probable mental retardation (Pape 2005: 210). Evidence collected by other experts suggests that "recruits who display signs of pathological behaviour are automatically weeded out for reasons of organizational security" (Taarnby 2003: 18). It seems reasonable to conclude that individualistic explanations based on psychopathology are of no value in helping us understand the rising incidence of suicide bombing in the world.

Deprivation

A second explanation of suicide bombing that focuses on the characteristics of perpetrators is the deprivation argument. In this view, suicide bombers act because they suffer extreme deprivation, either absolute or relative. **Absolute deprivation** refers to long-standing poverty and unemployment, **relative deprivation** to the growth of an intolerable gap between what people expect out of life and what they get (Gurr 1970). Presumably, deprivation of one sort or the other frustrates some categories of people until they are driven to commit self-destructive acts of aggression against the perceived source of their suffering.

Evidence does not support the deprivation theory. One researcher found education and income data on about 30 percent of Arab suicide bombers between 1980 and 2003. He reported that they were much better educated than the populations from which they were recruited. They were typically from the working and middle classes and were seldom unemployed or poor (Pape 2005: 213–15). Another scholar discovered that suicide bombers from Egypt and Saudi Arabia have come mainly from middle- or upper-middle-class families (Laqueur 2004: 16). The perpetrators of the September 11, 2001, attacks on the United States were well-educated, middle-class men. Such evidence lends no credence to the notion that suicide bombers are especially deprived in any absolute sense.

Arguments about relative deprivation are purely speculative. To date, no researcher has measured the degree to which suicide bombers are relatively deprived and compared their level of relative deprivation with that of non-suicide bombers. The consensus in the literature today is that suicide bombers do not experience extraordinarily high levels of deprivation, either relative or absolute (Stern 2003: 50–52; Taarnby 2003: 10–12; Victor 2003).

Culture

Explanations for suicide bombing that focus on the individual characteristics of the attackers began to fray in the late 1980s, partly because the evidence collected by researchers did not support them. Consequently, many analysts shifted their focus to the *collective* characteristics of suicide bombers and, in particular, to their culture.

Some social scientists attributed much of the collective violence in the world to a "clash of civilizations" between Islam and the West (Huntington 1996; for a critique, see Hunter 1998). From their point of view, Islamic culture inclines Muslims toward fanatic hatred of the West, violence, and, in the extreme case, suicide attacks. For example, the martyrdom of Hussein at the Battle of Karbala in 680 CE was a signal event in Islamic history, and it is often said to have reinforced the readiness of Muslims, especially Shi'ites, to sacrifice their lives for the collective good in the face of overwhelming odds (Reuter 2004: 37–39).

While such cultural resources likely increase the chance that some groups will engage in suicide attacks, one must be careful not to exaggerate their significance. One difficulty with the clash-of-civilizations thesis is that public opinion polls show that Arabs in the Middle East hold strongly favourable attitudes toward American science and technology, freedom and democracy, education, movies, and television, and largely favourable attitudes toward the American people. They hold strongly negative attitudes only toward American Middle East policy (Zogby 2002). This finding is less evidence of a clash of civilizations than of a deep political disagreement.

Nor is the notion that an affinity exists between Islam and suicide bombing supported by most Western students of Islam. According to one such expert, for example, "much of the so-called Islamic behavior that the West terms terrorism is outside the norms that Islam holds for political violence" (Silverman 2002: 91).

One must also bear in mind that secular Muslim groups in the Middle East and non-Muslim groups outside the Middle East have employed suicide bombing as a tactic. Of the 83 percent of suicide attackers between 1980 and 2003 for whom background data are available, 57 percent were non-religious (Pape 2005: 210). In Lebanon, Israel, the West Bank, and Gaza between 1981 and 2003, fewer than half of suicide missions were conducted by religious individuals (Ricolfi 2005), something that hardly increases confidence in the view that Islamic fundamentalism is the source of suicide bombing.

A final difficulty with cultural interpretations is that suicide attacks are by no means a constant in Islamic history. They appear in

11th-century northern Persia, in the 18th century in parts of India, Indonesia, and the Philippines, and in the late 20th century in various parts of the Muslim world. The episodic nature of suicide attacks suggests that certain social and political circumstances may be decisive in determining which cultural resources are drawn upon at a given time to formulate tactics for collective violence. For example, in the 18th century, suicide attacks were chosen as a tactic because little else proved effective against vastly militarily superior European and American colonial powers (Dale 1988). Similarly, militant Islamic groups in the late 20th and early 21st centuries adopted suicide bombing only after other tactics were tried and failed. Suicide bombing, it seems, is a weapon of last resort. All of this points to the difficulty of trying to explain political variables with cultural constants.

Focus on the Occupiers

Although cultural explanations of suicide bombing still have their supporters, a shift of focus occurred in the late 1990s. Scholars began to analyze suicide attacks as strategically rational political actions (Sprinzak 2000). With the publication of Robert Pape's study of all 462 suicide bombers who attacked targets worldwide between 1980 and 2003, this school of thought was given a strong empirical basis of support (Pape 2005).

According to Pape, every group mounting a suicide campaign since the early 1980s has shared one objective: to coerce a foreign state to remove its military forces from territory that group members view as their homeland (Pape 2005: 21). Pape makes his case by first quoting the leaders of organizations that have mounted suicide attacks. They stated plainly and forcefully that their chief aim is to liberate territory from what they regarded as foreign occupation or control (Pape 2005: 29–33). To support his claim that suicide bombing is a fundamentally rational strategy, Pape then notes that suicide attacks are not randomly distributed but occur in clusters as part of a campaign by an organized group to achieve a political goal. He identifies 18 suicide bombing campaigns that have taken place since the early 1980s, 5 of them ongoing (Pape 2005: 40).

Finally, Pape argues that the strategic rationality of suicide bombing is evident in the correlation between the increasing use of suicide bombing campaigns and their relative success in achieving their goals. He finds that suicide bombing has a roughly 50 percent success rate and regards that as high since, by comparison, international military and economic coercion achieves its goals less than a third of the time (Pape 2005: 65). In short, Pape claims that strategic rationality is evident in the *objectives*, *timing*, and *results* of suicide bombing campaigns.

Pape's research has convinced analysts that many instances of suicide bombing are not devoid of strategic logic. I contend, however, that it oversimplifies matters considerably to think that suicide bombing campaigns are launched only to liberate territory, that they are typically timed to maximize their impact in that regard, and that they often meet with success.

In the remainder of this chapter, I analyze one of the most protracted and destructive series of suicide bombings in the past quarter of a century—those of the second Palestinian *intifada* ("shaking off" or uprising) in Israel, the West Bank, and Gaza between 2000 and 2005. My analysis leads me to three conclusions:

1. With respect to *objectives*: Suicide bombing is an action that typically involves mixed motives and mixed organizational rationales. Strategic thinking is only one element that may combine with others in the creation of a suicide bomber. It predominates less frequently than Pape leads us to believe.
2. With respect to *timing*: Because the individual motivations and organizational rationales of suicide bombings are often mixed, suicide bombing campaigns are not always or even often timed to maximize the strategic advantages of insurgents. The timing of suicide bombings may be detached from strategic considerations because they take place for non-strategic reasons such as revenge or retaliation or simply when opportunities for attack emerge. As a result, their timing may not maximize the strategic gains of the attackers and, on occasion, may even minimize such gains.
3. With respect to *results*: Suicide bombing campaigns sometimes encourage targets to make minor concessions, but they often fail to achieve their main objectives. Sometimes, they have consequences that are the opposite of those intended by suicide attackers and their organizations. If suicide bombing

pays, as Pape claims, its net returns are often meagre and sometimes negative.

Suicide Bombing during the Second *Intifada*

One of my Ph.D. students and I collected information on all 138 suicide bombings that took place in Israel, the West Bank, and Gaza from October 26, 2000, to July 12, 2005 (Brym and Araj 2006). Our sources included the online database of the International Policy Institute for Counter-Terrorism (ICT) in Israel; the website of Israel's Ministry of Foreign Affairs; the East Coast evening edition of *The New York Times*; and two authoritative Arabic newspapers—*al-Quds*, published in Jerusalem, and *al-Quds al-'Arabi*, published in London. I was especially interested in three issues: (1) the reasons suicide bombers gave for their actions in public statements they made prior to attacking (that is, the bombers' *motives*); (2) the reasons that representatives of organizations claiming responsibility for suicide attacks gave for their actions (that is, the organizations' *rationales*); and (3) the preceding events that affected the timing of suicide bombings according to representatives of organizations claiming responsibility for the attacks (that is, the attacks' *precipitants*).

I classified the three causal mechanisms—bomber motives, organizational rationales, and event precipitants—as either "proactive" or "reactive." *Reactive* causes are Israeli actions that elicited a Palestinian reaction in the form of a suicide attack. Such Israeli actions include the assassination of organizational leaders and members, the killing of Palestinians other than organizational leaders and members, and other actions not involving killing, such as the demolition of houses owned by the families of people involved in anti-Israel activities. *Proactive* causes are political, religious, or ideological events that elicited a suicide attack without provocation by specific Israeli actions. In such cases, organizations used symbolically significant anniversaries, elections, or peace negotiations as opportunities to further their goals by means of suicide attacks.

I found that the great majority of suicide attacks during the second *intifada* were reactive, that is, provoked by specific Israeli actions (see Table 3.1). This finding has enormous implications for our understanding of the objectives, timing, and results of suicide bombing campaigns. Let us consider each of these issues in turn.

Table 3.1 Suicide Bombing and the Second *Intifada*: Causal Mechanisms (percentage)

Cause	Type		
	Reactive	**Proactive**	**Total**
Bomber motive (with implications for objectives)	71	30	101*
Organizational rationale (with implications for results)	59	41	100
Event precipitant (with implications for timing)	82	18	100

*Does not equal 100 because of rounding.

Source: Adapted from Brym and Araj (2006).

Objectives

Revenge and retaliation figured prominently in the bombers' stated reasons for planning suicide attacks. For the most part, they gave up their lives not as part of some grand rational strategy, but to avenge the killing of a close relative, as retribution for specific attacks against the Palestinian people, or as payback for perceived attacks against Islam. This finding supports the educated but impressionistic conclusion that Israeli political philosopher Avishai Margalit (2003) reached:

> Having talked to many Israelis and Palestinians who know something about the bombers, and having read and watched many of the bombers' statements, my distinct impression is that the main motive of many of the suicide bombers is revenge for acts committed by Israelis, a revenge that will be known and celebrated in the Islamic world.

Timing

Most suicide attacks were precipitated by Israeli actions. Their timing was, in that sense, not of the Palestinians' choosing and therefore not rationally planned to maximize strategic gains. To be sure, Israel's response to suicide bombings influenced the ease with which subsequent attacks could be mounted. Especially after the extraordinarily frequent and lethal suicide missions of March 2002, Israel's stepped-up counter-terrorist activities significantly decreased the number of suicide bombings and increased the time between precipitant and reactive attack. But I found little evidence to support Pape's contention that suicide attacks are timed to maximize the achievement of strategic or tactical goals. My analysis of precipitants leads me to conclude that most suicide bombings were revenge or retaliatory attacks and were advertised as such by insurgents.

Results

Pape's claim that suicide bombing achieves a relatively high rate of success in terms of reaching strategic goals is also questionable. Pape defines success as the withdrawal of occupying forces. The second *intifada* witnessed just one such withdrawal—Israel's August/September 2005 pullout from Gaza. Can the pullout be construed as a consequence of Palestinian suicide attacks? Two facts argue against such an interpretation.

First, when I examined the geographical locations of suicide bombings and the geographical origins of the bombers, I found that Gaza was the site of a disproportionately small number of suicide attacks and the recruiting ground for a disproportionately small number of suicide bombers. If suicide attacks were a decisive factor in leading to territorial concessions, one would expect those concessions to have been made not in Gaza but in the West Bank, where the great majority of bombers were recruited and from which the great majority of suicide attacks were launched (Brym and Maoz-Shai 2009).

Second, to the degree that militant Palestinian organizations mount suicide attacks to coerce Israel to abandon territory, the results of such attacks seem to be the opposite of what was intended. Rather than pushing the Israeli public to become more open to the idea of territorial concessions, suicide bombings have had the opposite effect. Israeli polls thus show that suicide attacks helped hardliner Ariel Sharon win the February 2001 election and, in general, drove Israeli public opinion to the political right throughout the second *intifada* (Arian 2001, 2002; Elran 2006).

Suicide bombings also encouraged Israel to reoccupy Palestinian population centres in the West Bank and Gaza. Israel had withdrawn from those population centres in 1995–97 as a result of peace talks. But in March 2002, 135 Israeli civilians were killed in suicide attacks, the most infamous of which was the Passover massacre at the Park Hotel in Netanya, in which 30 Israelis lost their lives. Within 24 hours, Israel launched Operation Defensive Shield. Twenty thousand reservists were called up in the biggest mobilization since the 1982 invasion of Lebanon and the biggest military operation in the West Bank and Gaza since the 1967 war. The West Bank and Gaza were almost completely reoccupied by April. Even if the strategic aim of the suicide bombings in March was purely to coerce Israel to withdraw from the occupied territories, the result of those attacks was just the opposite.

On a broader canvas, substantial West Bank territory has been incorporated on the Israeli side of the wall that Israel built to make it harder to launch suicide attacks. Therefore, in the long run, too, suicide bombings have made it more difficult for the Palestinians to gain territorial concessions from Israel. Many Palestinians themselves recognize that suicide bombing is a problematic strategy that rarely achieves strategic territorial goals and often has unintended, negative consequences from the Palestinian point of view. Among them is Palestinian President Mahmoud Abbas, who typically declares each suicide bombing "a crime against our people" ("al-Ra'is: ..." 2005).

Focus on the Interaction

How, then, can we explain the rise of suicide bombing since the early 1980s? There seems little advantage in focusing on the characteristics of suicide bombers themselves. On close inspection, neither their mental state nor their supposed deprivation nor their alleged cultural background adequately accounts for their actions.

Focusing on the occupier—seeing suicide bombing as a rational, strategic response to the perception that a foreign power has taken control of one's homeland—is a step forward, analytically speaking. The desire to regain control over territory does motivate some suicide bombers, accounts for the timing of some of their attacks, and sometimes results in concessions on the part of occupying forces. For example, my impression is that rational, strategic considerations linked to the desire to regain territory were more evident in the suicide bombings that took place in Israel, the West Bank, and Gaza between 1993 and 1997, and have been more evident in Iraq between 2003 and the present, than they were during the second *intifada* (Brym 2007).

However, my analysis of suicide bombing during the second *intifada* shows that focusing on occupation as the sole or even the most important reason for suicide bombings can be an oversimplification of a complex social process. In most of the 138 cases I examined, rational, strategic considerations linked to the desire to regain control of territory did not account for observable patterns in the objectives, timing, and results of suicide attacks.

It seems most fruitful to base explanations for patterns of suicide bombing on the *interaction* between occupied people and occupying forces. When other tactics fail to bring about strongly desired results, an occupied people may engage in suicide attacks out of desperation. However, a resolute occupier may have the will and the means to retaliate, often violently. Israel, for example, has responded to suicide attacks by engaging in the widespread assassination of Palestinian activists. Israel's actions have provoked more suicide attacks and other forms of collective violence on the part of the Palestinians, who are just as resolute as the Israelis, and renewed Palestinian violence has typically resulted in still more Israeli repression.

I conclude that patterns of collective violence, including suicide bombings, are not shaped by one side or the other in the conflict. They are governed by a deadly interaction—a lethal and escalating dialogue—between conflicting parties. Interpreting that dialogue is the sociologist's job.

AN ESCALATING DIALOGUE

Steven Spielberg's 2005 film, *Munich*, recounts the events surrounding the massacre of 11 Israeli athletes at the 1972 Olympics by Palestinian militants. A squad of Israeli secret service agents is quickly given the green light to track down and assassinate the Palestinians who masterminded the massacre. After the squad's first few hits, a letter bomb sent to the Israeli embassy in London kills an Israeli official. Other letter bombs are found at Israeli embassies in Argentina, Austria, Belgium, Canada, the Congo, and France. "They're talking to us," one Israeli agent says to another when he learns about the letter bombs. "We are in dialogue now."

The violent dialogue to which the secret agent refers has been going on for more than a century—ever since the first Eastern European Jews arrived in Palestine with the hope of establishing a Jewish homeland and Arabs objected to their presence (Mandel 1965). Periodically, each side in the conflict comes to the conclusion that an escalation in violence will finally silence the other side, but it never turns out that way. Instead, after a lull, renewed and more intense violence erupts.

The importance of each side's resolve cannot be underestimated in the perpetuation of the conflict. American and French troops abandoned Lebanon in 1983 after one suicide attack on each of their barracks. Spanish troops exited Iraq immediately after the Madrid train bombings in 2004. In these cases, relatively low resolve on the part of the perceived occupiers resulted in their making quick concessions to the attackers. In contrast, in the Israeli and Palestinian cases, the probability of serious concessions is low because the resolve on both sides is so high.

In the most recent phase of their battle over territory that both sides claim as their historical and religious birthright, one side was too weak to imagine a balance of power, so it concocted a scheme to achieve a balance of horror, justified by the idea that "a nation whose sons vie with each other for the sake of martyrdom does not know defeat" (quoted in Oliver and Steinberg 2005: 61). The powerful side responded to "martyrdom operations" (as suicide bombings are called by Palestinian militants) in the way that most of its enraged population demanded: by teaching the other side a series of lessons it wouldn't soon forget. The weak side obliged by remembering well and avenging its losses with all the fury it could muster.

Some of the thinkers in Israel's strategic planning offices surely recognized that murderous retribution is often counterproductive. They had to answer to their political bosses, however, who were, in turn, obliged to respond to public outrage by getting tough. Some of the Palestinian strategic thinkers in the warrens of Gaza City undoubtedly knew that Israel would not capitulate in response to suicide bombing. But they had to answer to their publics, too, and so they often forsook the strict calculation of costs and benefits for political expediency and a culture of mutual destruction.

To call the deadly interaction between Palestinians and Israelis "rational" distorts the meaning of the word. There is nothing rational about suicide bombing provoking assassination and assassination provoking more suicide bombing or other forms of lethal violence, such as rocket attacks. The interaction pushes both sides farther from their ultimate objectives of peace and security and threatens both sides with more horrible forms of violence in the future.

The irrationality of the interaction between Palestinians and Israelis was driven home to me at the New York conference I mentioned at the beginning of Chapter 2. It was a conference on human rights. One of the speakers was Dr. Yoram Dinstein, Israel's foremost expert on human rights law. Dr. Dinstein took part in a spirited debate on the legality of the Israeli policy of assassinating Palestinian militants. He reminded his audience that Israel is at war with a terrorist enemy, and Duchess of Queensberry rules therefore don't apply. He also suggested that Israel's assassination policy lowers the danger of violent acts against Israel by defusing human "ticking bombs."

I spoke to Dr. Dinstein after his lecture. "Legal issues aside," I asked, "do you really think that Israel's policy of targeted killings is rational?" I proceeded to tell him about my research suggesting that the assassination of Palestinian militants provokes more suicide attacks and other forms of lethal violence. I also argued that assassinations help radicalize Palestinians, making it easier to recruit a new, larger, more determined, and more ruthless generation of militants. Finally, I mentioned the collaborator problem. Assassinations require real-time information on the whereabouts of targets. A large network of Palestinian collaborators feeds this information to the Israeli security services. The existence of this network causes mistrust, conflict, and internal violence among Palestinians in the West Bank and Gaza. Such social chaos undermines the unity and stability of Palestinian society that is required if one wants a negotiating partner who can make binding, authoritative decisions (Gross 2003). Maintaining a wide network of Palestinian collaborators not only helps the Israeli security services locate targets in real time, but it also helps undermine whatever slight chance for peace remains in the region. I concluded that, regardless of their legal status, targeted killings are politically irrational: They are intended to stop violence but have the effect of perpetuating hostility.

Dr. Dinstein dismissed my claims with a wave of the hand.

"So," I suggested, "you conclude that the Palestinians understand only force?" To which he replied, "Even that they don't understand."

And he was right. If the Palestinians understood force, they would have capitulated long ago. Instead, the exercise of repressive force by Israel only deepens their resolve. This observation raises two obvious questions that the legions of legal and counterterrorist experts have been unable to answer: If the Palestinians don't respond to the use of repressive force as "reasonable" people ought to, what is the good of using it? And if Israelis don't make concessions as Robert Pape says they should, what is the good of launching suicide attacks against them?

SOCIAL INTERACTION

Max Weber, a founding father of sociology, defined **social action** as human behaviour that is meaningful in the sense that it takes into account the behaviour of others. From his point of view, one person may intervene in a situation, a second may deliberately refrain from intervention, and a third may passively acquiesce in the situation. But all three act socially if their behaviour results from taking into account what others are likely to do. Tripping on a rock is not a social action, but failing to speak up for fear of punishment is (Weber 1947: 88).

Social interaction is a dynamic sequence of social actions in which people (or entire categories of people) creatively react to each other. Social interaction is of such fundamental importance that, without it, individuals would not be able to develop a sense of identity, an idea of who they are. Nor would a mere social category (such as the residents of a particular street) be able to crystallize into a self-conscious social group (such as a true neighbourhood).

Individual and group identity formation is possible only because humans enjoy a highly developed capacity to empathize or "take the role of the other" (Mead 1934). We develop a sense of who we are by interpreting the actions of others and imagining how they see us. All social interaction, including interaction that involves conflict, sharpens one's identity. In fact, nothing makes people feel more a part of their nation than a good war (Coser 1956: 87–103).

The capacity to take the role of the other is especially valuable in conflict situations because it increases the likelihood of conflict resolution. This truth is well illustrated by *The Fog of War*, which won the 2003 Oscar for best documentary film. The film surveys the life of Robert McNamara, U.S. Secretary of Defense during the Kennedy and Johnson administrations and architect of the Vietnam War. In the film, McNamara outlines 11 lessons that he learned over his years of public service. His lesson number one: Empathize with your enemy.

McNamara was present in October 1962 when President Kennedy was ready to start a nuclear war with the Soviets if they didn't remove their missiles from Cuba. Kennedy believed that Khrushchev, the Soviet leader, would never negotiate a removal. But Llewellyn "Tommy" Thompson, former U.S. ambassador to Moscow, disagreed. Thompson knew Khrushchev personally and understood that he would back down from his belligerent position if presented with an option that would allow him to remove the missiles and still say to his hard-line generals that he had won the confrontation with the United States. "The important thing for Khrushchev," Thompson argued, "is to be able to say, 'I saved Cuba; I stopped the invasion.'" Thompson convinced Kennedy. Negotiations began, and nuclear war was averted.

"That's what I call empathy," McNamara observes. "We must try to put ourselves inside [the enemy's] skin and look at us through their eyes." Note that being empathic does not mean having warm and fuzzy feelings about an enemy. It means understanding things from the enemy's perspective in order to design a resolution that will enable the greatest gains and the fewest losses.

The great tragedy of the Israeli–Palestinian conflict is that it has been so bitter and protracted that the capacity of each side to empathize with the other has been deeply eroded. An increasingly large number of Israelis believe that the Palestinians want to destroy Israel as a Jewish state, and an increasingly large number of Palestinians believe that the Israelis want to prevent the creation of a viable Palestinian state. More and more, Palestinians fail to appreciate the legitimate security needs of Israel, and Israelis fail to appreciate the legitimate national ambitions of the Palestinians. No Nelson Mandela-like figure who can peacefully reconcile the warring parties has risen above the fray, and in recent years the United States has not shown much willingness to drag both sides to the negotiating table and use its political and economic might to compel them to hammer out a resolution. It is, therefore, unclear whether the impasse can be broken anytime soon.

CRITICAL THINKING EXERCISES

1. According to this chapter, what is the main condition needed for bringing an end to violent conflict? Do you agree or disagree with this assessment? On what grounds? How relevant is this condition for bringing an end to conflict in general, whether between individuals, groups, or nations? Answer these questions in three pages.

2. *Paradise Now* is a film about Palestinian suicide bombers. It was nominated for the 2005 best foreign film Oscar. In the movie, best friends decide to undertake a suicide mission. View the movie. In a page, explain why one friend decides to go through with the mission while the other doesn't. Imagine yourself in the place of each friend. In a page, explain why you would or would not act as that person did.

Aaron Millard

4

Hurricane Katrina and the Myth of Natural Disasters

A RACIST PRESIDENT OR AN ACT OF GOD?

On September 2, 2005, the NBC television network broadcast a telethon in support of American Red Cross disaster relief efforts along the coast of the Gulf of Mexico. Hurricane Katrina, one of the largest hurricanes of its strength ever to reach the United States, had made landfall about 80 hours earlier. Large swaths of Louisiana, Mississippi, and Alabama lay flooded and in ruins. After Harry Connick, Jr., sang "Do You Know What It Means to Miss New Orleans?" Canadian comedian Mike Myers and rapper Kanye West took the floor. Myers faithfully followed the teleprompter and described the wretched state of New Orleans and its people. West, however, veered wildly off script. He damned the mass media for their portrayal of black people as looters, criticized the government for taking so long to arrive with aid, and concluded with the memorable sentence, "George Bush doesn't care about black people." Myers then asked viewers to "Please call ..." but didn't get to finish his sentence. Someone in the NBC control room apparently figured out where West had been headed and ordered the camera to turn away and cut to comedian Chris Tucker (Dyson 2006: 26–27).

Two months later, hip hop star 50 Cent was interviewed by Contactmusic.com. "I don't know where that came from," he said, referring to Kanye West's televised outburst. "The New Orleans disaster was meant to happen. It was an act of God" ("50 Cent Slams" 2005).

The comments by Kanye West and 50 Cent received a lot of attention and provoked much debate over whether the disaster was the result of one man's alleged racism or nature's wrath. From a sociological point of view, however, neither rap star came close to understanding how it came about that a storm in the world's richest and most powerful country could kill 2300 people, cause more than US$100 billion in damage, seriously disrupt the supply of oil and natural gas to the nation, and force the eventual evacuation of 80 percent of New Orleans' population.[1] After all, the danger of such a storm was widely and precisely known years earlier. The Federal Emergency Management Agency (FEMA) issued a report in early 2001 saying that a hurricane striking New Orleans was one of the three most likely disasters to hit the United States (the others were a terrorist attack on New York City and a major earthquake in San Francisco). Since 2001, long, detailed, and, as Hurricane Katrina later proved, shockingly accurate articles had appeared in *Scientific American*, *Time*, *National Geographic* magazine, *Popular Mechanics*, *The New York Times*, and New Orleans' *The Times-Picayune* that made the results of research on the effect of a powerful hurricane hitting New Orleans available to the broad public and its political representatives (see, for example, "Washing Away" 2002). Yet almost nothing was done to prepare for the inevitable.

Explaining this sociological mystery is the chief aim of this chapter. My explanation consists of two main parts. First, for centuries, powerful and well-to-do people made economic and political decisions that placed New Orleans, and especially its poor black citizens, at high risk of hurricane-related death. Second, for an equally long period, powerful and well-to-do people resisted charging the American government with responsibility for ensuring the welfare of the citizenry as a whole. As a result, relatively inexpensive measures that prevent hurricane-related deaths in other countries have not been implemented in the United

[1] In June 2005, the Gulf of Mexico was responsible for nearly 30 percent of U.S. oil production and 20 percent of U.S. natural gas production (Energy Information Administration 2005). In addition to 1836 confirmed deaths due to Katrina as of mid-May 2006, experts estimate that roughly 500 Louisiana residents were swept away and will never be found or identified (Krupa 2006).

States. Neither God nor one man should be held responsible for the decisions and neglect of entire social classes.[2]

THE DEVELOPMENT OF NEW ORLEANS

In 1840, New Orleans was the fourth most populous city in the United States, and until the 1920s, it was the world centre of jazz. On the eve of Hurricane Katrina, it was still an important port and tourist town, with a metropolitan population of more than 1.3 million. And, of course, it had a reputation. Tennessee Williams's *A Streetcar Named Desire* branded New Orleans sensual and decaying. John Kennedy Toole's *A Confederacy of Dunces* rendered it a magnet for loose screwballs. Anne Rice added to its mystery in *Interview with the Vampire*. Everyone knew it as a party town, home of the Mardi Gras, a place that gave the world gumbo and jambalaya, and the only city in North America with a major street named after a 90-proof liquor.

New Orleans is situated on the coast of the Gulf of Mexico (see Figure 4.1). Most of it lies below sea level—in places, as much as 2.5 m (8 feet) below. To the south, the Mississippi River flows past the city, through wetlands, and into the Gulf. To the north lies Lake Pontchartrain, the second-biggest saltwater lake in the United States and the largest lake in Louisiana (see Figure 4.2). Imagine half a dozen exuberant eight-year-olds splashing in a swimming pool on a hot summer afternoon. New Orleans is like a plastic soup bowl floating in the pool.

About 1.6 m (5.3 feet) of rain falls on New Orleans annually—more than 2.3 times the annual rainfall in Toronto. Every spring, the Mississippi tries to flood, and in the past, it often succeeded, even after people began building dykes (or "levees," as they are called locally) to block the overflow. In the 20th century, city residents achieved a measure of control over flooding by constructing a series of canals that allow 900 million cubic metres (24 billion gallons) of water to be collected and pumped into Lake Pontchartrain and other nearby bodies of water every day. Still, every summer and fall, tropical storms and hurricanes assault the Gulf Coast. Sometimes, great waves of seawater surge into New Orleans. Levees were built to protect the city from storm surge, too. Yet there is enough threatening water in the area to make a

[2] **Social class** is one of the most important concepts in sociology and much controversy surrounds its definition. For my purposes, it is sufficient to define social class as a position occupied by people in a hierarchy that is shaped by economic criteria, including wealth.

Figure 4.1 The Caribbean Basin and Gulf Coast

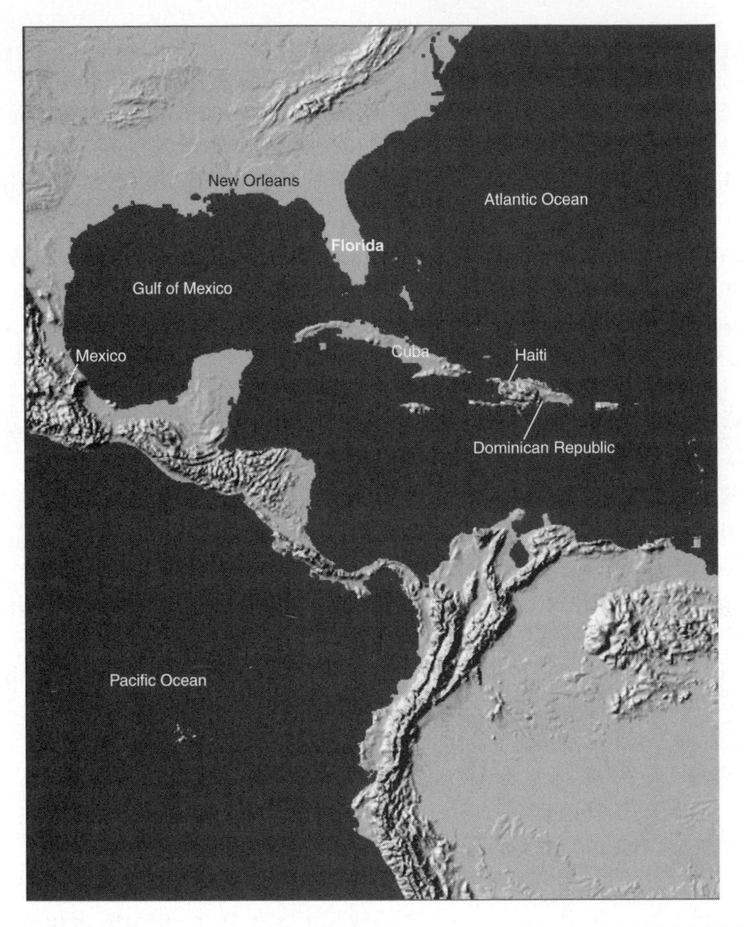

Source: Adapted from U.S. Geological Survey, Department of the Interior/USGS (2006).

reasonable person ask why the French ever settled New Orleans in the first place.

They did so as part of their strategy for continental control. The Mississippi leads deep into the North American interior, links to other rivers that empty into the Great Lakes, and thus offers access to Canada, which the French had founded in 1608 (calling it "New France"; Sexton and Delehanty 1993). The swampy and treacherous Mississippi Delta was difficult for the French to negotiate, so they settled 200 km (125 miles) upstream from the river mouth.

Figure 4.2 Hurricane Katrina Disaster Areas

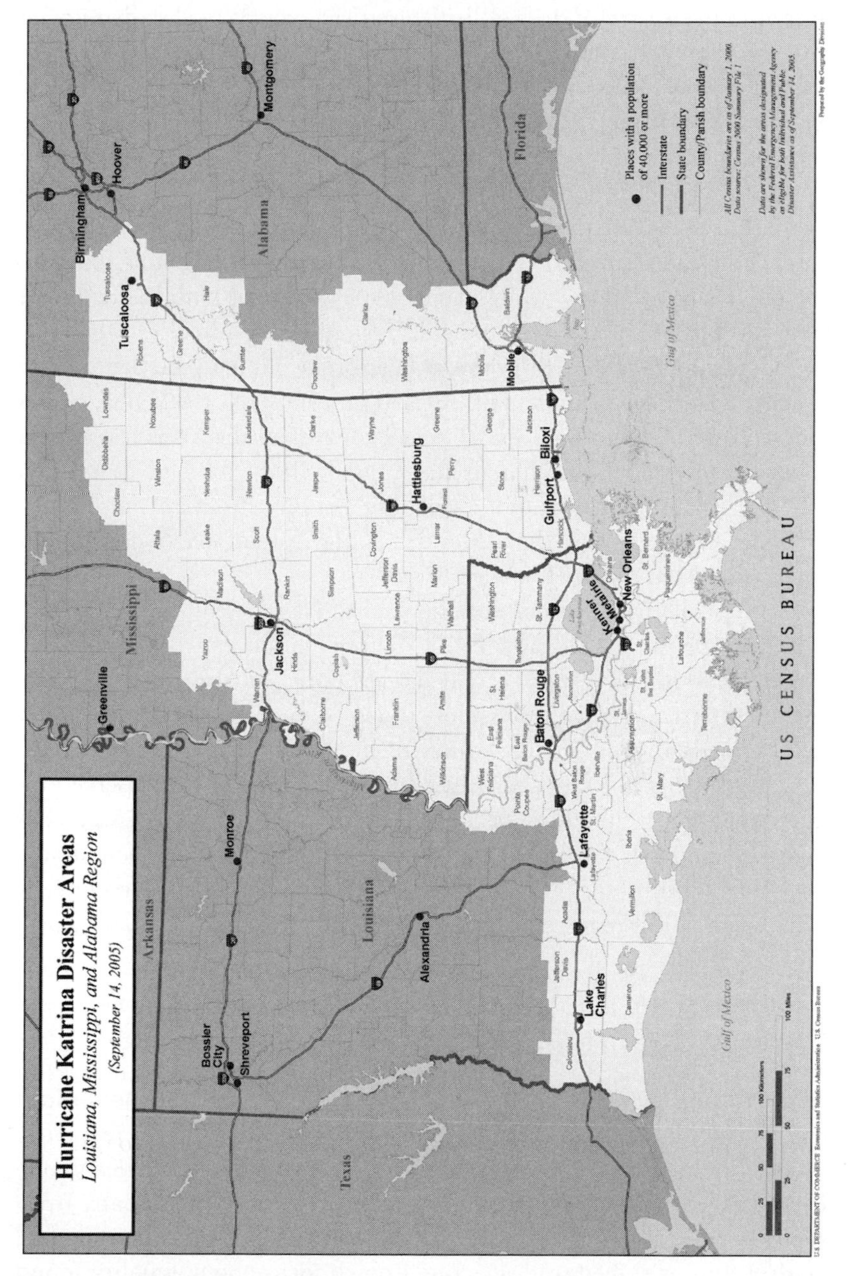

Source: U.S. Census Bureau (2005).

Thus was New Orleans born in 1718. It served as one end of a continental bridge between French land holdings, a bridge that also served as a barrier to the westward drift of British settlers, who were already moving into the disputed Ohio River Valley. Soon, several thousand French, Canadian, and German colonists, West African slaves, and natives were living in New Orleans. Mainly because women were scarce, intermarriage was common, resulting in the formation of a distinct ethnic group, the Louisiana Creole, a social jambalaya that was unusually open to all races and cultures. With typical colour and exaggeration, yet with a grain of truth, Louisiana Governor Huey Long said two centuries later that you could feed all the "pure" white people in New Orleans with half a cup of beans and half a cup of rice, and still have food left over.

In the middle of the 18th century, war broke out between Britain and France. France lost Canada and its grip on Louisiana weakened. Spain took control of Louisiana for 38 years, and when the French resumed control, they sold Louisiana to the United States for US$15 million (about US$400 billion in today's dollars), in one stroke raising money for their next war with Britain and helping to reinforce a power that could rival France's chief enemy.

White American settlers now flocked to New Orleans. Census data show that between 1810 and 1860, the population grew tenfold as the city became the country's second-largest port. Tobacco, lumber, rice, sugar, cotton, and grain were shipped out. Manufactured goods, slaves, luxury goods, and coffee flowed in. In that half century, the number of New Orleanians of European origin increased from 37 to 85 percent of the population, while those of African origin fell from 63 to 15 percent (Logsdon and Bell 1992: 206). A distinct, modern American neighbourhood was built uptown, its inhabitants separated from the Creole residents of the old French city by broad Canal Street and the Americans' sense of social, moral, and economic superiority.

The Growth of Black–White Inequality

New Orleans before the Louisiana Purchase was not a city without racial inequality. Black slaves were brought there from the beginning. Still, the colour line was less rigid in New Orleans than in the cities of the United States. Racial intermarriage was relatively common. Many freed slaves lived in New Orleans, and they were often employed not as menial workers but as skilled tradespeople. French ideas about equality found eager supporters in the city. As a result, no sharp line separated blacks from whites. Social contact flourished among all categories of New Orleanians, fostered by musicians, live-in lovers, the Catholic clergy, grocers, and saloon keepers. Class, culture, and complexion elevated many black residents to a status that most Americans of African descent could envy but not reach (Logsdon and Bell 1992).

Once Americans started settling in the city, the situation of the black population deteriorated. In the 1840s and 1850s, an influx of white American workers displaced many freed slaves who worked on the docks and in skilled trades. Racism increased as job competition mounted; some freed slaves left for Haiti to escape the prejudice and discrimination. Louisiana's fight to protect slavery during the Civil War (1861–65) hardened the colour line. Then, from 1890 to 1952, a series of laws institutionalized racial segregation. Blacks and whites now had to be kept apart in trains, schools, streetcars, bars, prisons, homes for elderly people, and even in circus audiences. They could not marry or cohabit, nor could they adopt a child of a different race. They could not dance or engage in athletic competition together. Blacks could not receive permits to build houses in white neighbourhoods, and whites could not receive permits to build houses in black neighbourhoods ("Jim Crow Laws" 2006). A comparison of the situation before and after the Louisiana Purchase illustrates the fact that **races** are defined not so much by biological differences as by social forces: Racial distinctions are typically made and reinforced by advantaged people for the purpose of creating and maintaining a system of inequality.

In the 1920s, new levee construction permitted the shoreline of Lake Pontchartrain to be heightened and extended, creating desirable real estate north of the city for white residents (Hirsch and Logsdon 1992). After World War II, new highways accelerated suburban growth. As a result of "white flight" to the suburbs, African Americans became a majority in the city of New Orleans proper after 1980, a position they had not held for 140 years. Just before Katrina hit, the city of New Orleans was 68 percent African American and just 25 percent non-Hispanic white. The poverty rate for African Americans was 35 percent—7.5 percent above the national average for African Americans. Two-thirds of the city's public schools were deemed "academically unacceptable" by the U.S. Department of Education. The city's homicide rate was the highest of any city in the country (Mahoney and Freeman 2005). And the black population was increasingly concentrated in the city's least desirable, low-lying areas.

Flood Control?

The situation of New Orleanians was especially precarious because flood control measures were inadequate (Blumenthal 2005; Bourne 2004; Nordheimer 2002; Tidwell 2004). Two big problems existed:

1. *The disappearance of coastal wetlands.* When waves of seawater from the Gulf of Mexico are whipped up by hurricane-force winds, they pound New Orleans. The first line of defence against this onslaught consists of the marshes and barrier islands of the wetlands between New Orleans and the Gulf. Every 3.2 km (2 miles) of wetland reduces storm surge by 15 cm (6 inches). Before levees were built along the Mississippi, silt from the river's floodwaters used to stop or at least slow down the sinking of coastal wetlands into the Gulf. But the levees divert silt into the Gulf, causing the wetlands to disappear at an alarming rate. The first line of defence against storm surge has thus been weakened.

   In addition to removing much of the physical barrier against storm surge, the disappearance of the wetlands has another negative consequence for flood control. Hurricanes are machines fuelled by heat. They gain strength from the heat that is released as vapour from warm seawater condenses and falls as rain. They weaken as they pass over land, which is cooler and, of course, drier. The disappearance of coastal wetlands has effectively brought the warm waters of the Gulf closer to New Orleans, ensuring that hurricanes have less chance to weaken before they make landfall.
2. *The inadequacy of the levees.* The second big flood control problem is that the levees along Lake Pontchartrain and in other areas around the city were last reinforced with higher walls after Hurricane Betsy struck in 1965, killing more than 70 people. The U.S. Army Corps of Engineers then built up the storm walls to withstand a Category 3 storm.[3] Forty years after Betsy, the Lake Pontchartrain and other levees desperately required upgrading, as

[3] The Saffir–Simpson scale ranks hurricanes as follows (National Weather Service, 2005):

- Category 1: Minimum one-minute sustained winds of 74–95 mph (119–153 km/hr) and above-normal storm surge of 4–5 ft (1.2–1.5 m)
- Category 2: Minimum one-minute sustained winds of 96–110 mph (154–177 km/hr) and above-normal storm surge of 6–8 ft (1.8–2.4 m)
- Category 3: Minimum one-minute sustained winds of 111–130 mph (178–209 km/hr) and above-normal storm surge of 9–12 ft (2.7–3.7 m)
- Category 4: Minimum one-minute sustained winds of 131–155 mph (210–249 km/hr) and above-normal storm surge of 13–18 ft (4.0–5.5 m)

> Hurricane Katrina painfully demonstrated; Katrina made landfall in Louisiana as a Category 3 storm, and the surge off the lake smashed the levees, engulfing the city. In all, about half the levee system was damaged (Burdeau 2006). What is worse, in any given year, New Orleans stands an estimated 1 percent chance of facing a Category 5 hurricane. Leaving the levees at less than Category 3 readiness was like playing Russian roulette using an atom bomb instead of a bullet.[4]

Aware of the problems just outlined, the federal government acted, but without resolve. It established a task force to help restore lost wetlands around New Orleans in 1990. In 2003, however, the Bush administration effectively ended that effort by allowing largely unrestricted development in the wetlands. Remarkably, in November 2005, two months *after* Katrina, the Bush administration refused to fund a US$14 billion plan to restore the barrier islands and wetlands. (If that seems like a lot of money, bear in mind that it equalled just six weeks of spending for the war in Iraq or 7 percent of the estimated cost of restoration following Katrina; Tidwell 2005.)

In addition, Congress authorized a project to improve the pumping of water out of the New Orleans area in 1996. Unfortunately, the project was only half finished when money effectively dried up in 2003. The Corps of Army Engineers received money to improve the levees on Lake Pontchartrain and vicinity. The Bush administration, however, cut funding for the project by more than 80 percent in 2004 and made additional cuts at the beginning of 2005, ranking other priorities, such as the war in Iraq, higher. Because of budget cuts, the Corps was unable to buttress the 17th Street levee on Lake Pontchartrain, the location of the biggest levee breach during Katrina.

BREWING STORMS

New Orleans on the eve of Katrina was unusually poor, black, segregated, unequal, violent, and vulnerable to flooding. As we have seen, there was nothing natural about this state of affairs. Centuries of human effort—in the form of geopolitical rivalry, economic competition, public policy, and social exclusion—were required to create it.

[4] Compiled from United States data (2004b: 38, 146). A Global Report Reducing Disaster Risk: A Challenge for Development. United Nations Development Programme. Bureau for Crisis Prevention and Recovery, www.undp.org/bcpr.

People's actions may have contributed to New Orleans' vulnerability in another way, too. I refer to the increasing use of fossil fuels such as gasoline, oil, and coal. Burning fossil fuel releases carbon dioxide into the atmosphere. Carbon dioxide is a heat-trapping gas; it allows more radiation to enter the atmosphere than escape it. The result is global warming. In turn, global warming may increase the intensity of tropical storms.

I say "may" because controversy surrounds the last part of my argument. Hardly any climate scientists doubt that the atmosphere is heating up or that the concentration of carbon dioxide and other heat-trapping gases has increased since the Industrial Revolution (Goddard Institute for Space Studies 2016). A large majority of climate scientists believe there is a cause-and-effect relationship at work here, not a coincidence.[5] Leading scientific bodies in the United States, including the National Academy of Sciences, the American Meteorological Society, the American Geophysical Union, and the American Association for the Advancement of Science, agree that the evidence for human impact on climate is compelling.

A study of 928 papers on climate change published in scientific journals between 1993 and 2003 found not a single one that disagreed with the consensus view (Oreskes 2004). In 2007, the report of the Intergovernmental Panel on Climate Change, which was written by 150 leading experts from more than 30 countries and reviewed by more than 600 scientific authorities, concluded that "human activities ... are responsible for most of the warming observed over the past 50 years" (Intergovernmental Panel on Climate Change 2007: 97). True, some scientists dispute the consensus. But their criticism should be taken with a grain of salt because much of their research is funded by the coal and petrochemical industries.

How might global warming affect hurricanes? In brief, global warming causes more water to evaporate. More vapour in the atmosphere may increase the frequency of hurricanes and cause the hurricane season to start earlier (Emanuel 2005; Holland and Webster 2007; Knutson and Tuleya 2004; Webster, Holland, Curry, and Chang 2005). Some scientists question whether enough data have yet been collected to substantiate

[5] More precisely, they believe that there is a cause-and-effect relationship with feedbacks that accelerate disequilibrium. For example, global warming melts permafrost. When permafrost melts, it releases methane, a much more potent heat-trapping gas than carbon dioxide. Global warming also melts the polar ice caps. When white ice is turned into dark ocean water, more solar radiation is absorbed by the earth and less is reflected back into space. Through these positive feedback loops, small temperature changes cause bigger temperature changes (Kolbert 2006).

these findings (Schiermeier 2005a, 2005b). Therefore, the link between global warming and hurricane destructiveness must still be treated as an intriguing although strong possibility rather than as a proven fact.

One thing can be said with certainty, however: If research substantiates the connection, it will be a global problem with deep local roots. With 4.6 percent of the world's population, the United States is responsible for 21.4 percent of the world's carbon dioxide emissions (Germanwatch 2007: 11).

A COMPARATIVE PERSPECTIVE

The strongest argument that deaths due to hurricanes are more a social than a natural disaster comes not from climate science but from sociology. It is an argument in three parts:

1. The populations of some countries are more exposed to the threat of hurricanes than are the populations of other countries.
2. *At the same level of exposure*, some countries experience relatively few deaths due to hurricanes while others experience relatively many such deaths.
3. Countries that experience relatively few deaths take extensive precautions to avoid the catastrophic effects of hurricanes. Countries that experience relatively many deaths take few such precautions.

Figure 4.3 adds weight to this argument. The graph contains data from 34 countries that were exposed to hurricanes from 1980 to 2000. It plots the number of people in each country who were exposed to hurricanes (along the horizontal axis) against the average number of deaths due to hurricanes each year (along the vertical axis).[6] In the period 1980–2000, the number of people exposed to hurricanes ranged from just over 18 000 in the small West African country of Cape Verde to more than 579 million in China. Average annual deaths due to hurricanes ranged from less than 0.5 in New Zealand to more than 7400 in Bangladesh.

In general, countries with large exposed populations experienced more annual hurricane-related deaths than do those with small exposed populations. That tendency is illustrated by the upward-sloping trend line. The line shows the number of deaths one would expect in an exposed population of a given size. Countries lying above the line were more vulnerable to hurricane-related deaths than one would expect given the size of their

[6] The data in Figure 4.3 are logarithms.

Figure 4.3 Relative Vulnerability to Hurricanes, 1980–2000

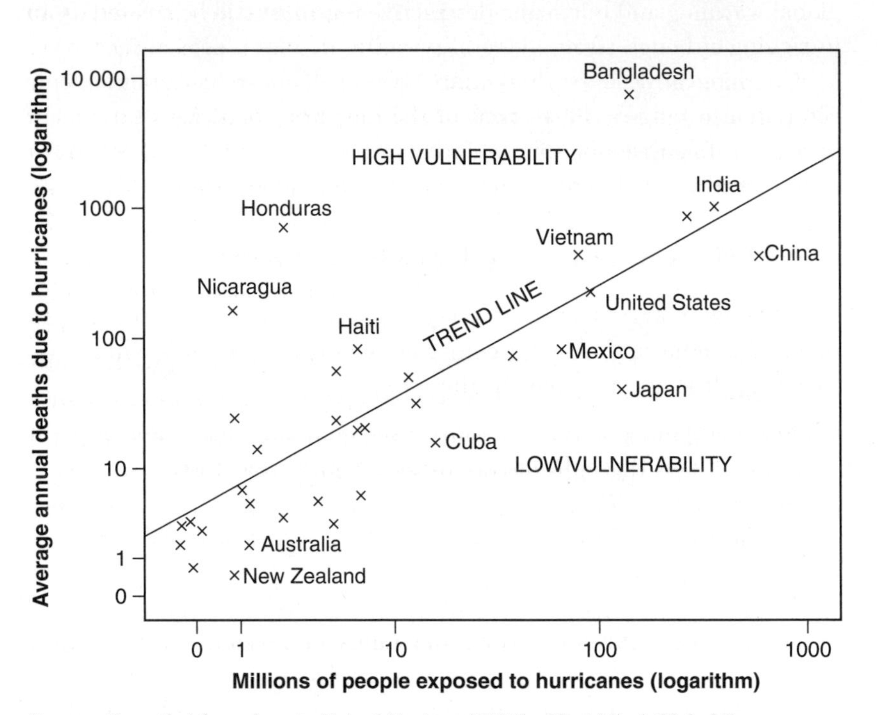

Source: Compiled from data in United Nations (2004b: 38, 146). A Global Report Reducing Disaster Risk: A Challenge for Development. United Nations Development Programme. Bureau for Crisis Prevention and Recovery, www.undp.org/bcpr.

exposed population. Countries lying below the line were less vulnerable than one would expect given the size of their exposed population.

This is where the graph starts to get interesting. The exposed populations of Bangladesh and Japan were approximately the same size (126 million in Japan, 135 million in Bangladesh). Based on the size of their exposed populations, one would expect both countries to have experienced an average of about 250 hurricane-related deaths per year. Yet Japan experienced an average of only 39 hurricane-related deaths per year, while Bangladesh experienced more than 7400. That is partly because Japan took far more extensive precautions to guard against such deaths than Bangladesh did.

True, Bangladesh is one of the world's poorest countries while Japan is one of the richest. The Japanese can, therefore, afford to take precautions that the Bangladeshis can only dream of. But wealth is not the only factor that determines a country's ability to take precautions. After all, the United States is about as wealthy as Japan is and had a considerably smaller exposed population (89 million people), yet it experienced an average of 222 hurricane-related deaths per year compared to Japan's 39. In fact, of the four rich countries represented in Figure 4.3, three of them—Japan, Australia, and New Zealand—experienced considerably fewer hurricane-related deaths than one would expect given the size of their exposed populations. But the fourth rich country—the United States—experienced about as many hurricane-related deaths per year as a relatively poor country with an exposed population of the same size. In the period 1980–2000, Americans were more vulnerable to death by hurricane than Mexicans were (with 65 million exposed people and an average of 85 hurricane-related deaths per year), and much more vulnerable than Cubans (with 11 million exposed people and an average of just three hurricane-related deaths per year).

Cuba

The Cuban case illustrates well the kinds of precautions a country can take to prevent hurricane-related deaths, even if it is relatively poor (Cohn 2005; Hamilton, de Valle, and Robles 2005; Martin 2005; Reuters News Agency 2005; United Nations 2004a). The last time Cuba suffered a large number of casualties because of a hurricane was 1963, when Hurricane Flora killed 1126 people. After Flora, the Cubans took action to prevent a recurrence of the tragedy.

Cuba first implemented an educational program in schools, universities, and workplaces to teach people how to prepare for and cope with natural disasters. The program trains the population from a young age to interpret and heed weather reports, which are broadcast continuously and updated frequently in the event of an impending storm. Every May, an annual two-day training session known as "Meteoro" is held. It focuses on risk reduction, including exercises that simulate evacuation and rescue in the event of a hurricane. Meteoro also encourages preventive measures, such as trimming tree limbs and checking for weaknesses in dams, and it involves a review and update of all emergency plans in the light of what has been learned in the preceding year.

Cuba has organized its civil defence network to coordinate evacuation and rescue operations at the neighbourhood level in the event of a big storm. Civil defence workers and members of organizations such as the Federation of Cuban Women go door to door to ensure that people fill their bathtubs with water, tape their windows, put their cars in the garage, unplug electrical appliances, and have an adequate supply of batteries, transistor radios, candles, matches, drinking water, and food. They create lists of the ill, elderly people, people with disabilities, and single mothers—people who would need help evacuating—and they ensure that help is available if these people are required to leave their homes.

If evacuation is necessary, neighbourhood doctors evacuate together with residents so people who need medication can be properly treated. Refrigerators, TV sets, pets, and other valuable items are evacuated with the people so they won't be reluctant to leave. Evacuation routes, means of transportation, and temporary housing facilities for evacuees (mainly in schools) are set up well in advance, as are stores of emergency water, food, and medicine. The existence and locations of these stores are widely publicized. Buses, trucks, ambulances, vans, helicopters, even horse carts are mobilized to get people to shelter. Regular water supplies are turned off to avoid the spread of disease.

As a result of these precautions, Cuba lost only 22 lives in 10 major hurricanes between 1985 and 2004. But the big test for the Cuban system of preventing hurricane-related deaths came between July 7 and 9 in 2005, when Hurricane Dennis, the most ferocious storm since Flora, lashed the island. Dennis hit Cuba twice as a Category 4 storm (Katrina hit Louisiana once as a Category 3 storm). Some 120 000 houses were badly damaged, 2.5 million people were left without electricity, and 12 000 hectares (30 000 acres) of banana trees were flattened. However, the timely evacuation of 1.5 million people—a remarkable 13 percent of Cuba's entire population—minimized the loss of life. Dennis killed just 16 Cubans. One can crudely estimate that if precautions similar to those employed in Cuba had been taken in the United States nearly three months later when Katrina hit, the American death toll would have been 128 people rather than 2300.[7]

Cuba is a communist country. Some people might argue that its sterling achievement in preventing hurricane-related deaths has been

[7] My logic is as follows: The United States is about eight times more exposed to hurricanes than Cuba is (89 million vs. 11 million exposed people). There were 16 Cuban deaths due to Dennis, a storm of roughly the same magnitude as Katrina. It follows that, if the United States had taken precautions similar to Cuba's, Katrina would have killed approximately 128 Americans (since $16 \div 11\,000\,000 = 128 \div 89\,000\,000$).

accomplished only by means of strict political control of its population—control that freedom-loving Americans would never tolerate. But the argument is suspect. Vietnam is also a communist country with strict political control of its population, yet it is highly *vulnerable* to hurricane-related deaths (see Figure 4.3). Japan is a capitalist country *without* strict political control of its population, yet its record of preventing hurricane-related deaths is better than Cuba's. Communism versus capitalism is not the issue here. What is decisive in determining a country's tolerance of hurricane-related deaths is its population's collective will to take responsibility for helping citizens who are in need, a will that is typically expressed through government policy. Compared to other rich countries, that collective will is weak in the United States.

KATRINA

Here is how it went down. On August 25, 2005, Hurricane Katrina hit Florida as a Category 1 storm, killing seven people.[8] It then veered into the Gulf of Mexico, where it was soon upgraded to Category 3. Kathleen Blanco, governor of Louisiana, declared a state of emergency on August 26, and President Bush followed suit on the 27th. New Orleans mayor Ray Nagin declared a voluntary evacuation order that night. At 10 a.m. on the 28th, Nagin finally announced a mandatory evacuation order. At that point, Katrina was a Category 4 storm about 20 hours from landfall. Seventeen hours later, at 3 a.m. on August 29, the 17th Street levee on Lake Pontchartrain collapsed. Other major levee breaches occurred at the London Avenue canal and the Industrial canal. Eighty percent of New Orleans was soon flooded, in some places to a depth of nearly 5 m (16 feet).

The response of government to the impending disaster was remarkably restrained, to put it politely. Top officials gravely underestimated the severity of the catastrophe about to befall residents of the Gulf Coast. Mayor Nagin hesitated to call a mandatory evacuation order because he was worried that the city would be legally liable for closing hotels, hospitals, and businesses. Consequently, he turned down an offer by Amtrak to evacuate several hundred New Orleanians on the last train out of town, and he failed to mobilize the city's 804 operational buses to

[8] The following account is based mainly on Dyson (2006), U.S. House of Representatives (2005), and reports in *The New York Times*.

get people out. The director of the National Hurricane Center had to call the mayor at home during dinner on the evening of the August 27 to tell him that the storm was the worst he had ever seen and practically beg him to declare a mandatory evacuation order the next morning.

The White House found out about the 17th Street levee breach at midnight on August 29, but on the morning of the 30th, President Bush, vacationing at his Texas ranch, expressed relief that New Orleans had "dodged the bullet." An e-mail confirming the levee breach had arrived at the Department of Homeland Security two-and-a-half hours before the information reached the White House, but the next morning, Michael Chertoff, Secretary of the Department of Homeland Security, flew to Atlanta for a conference on bird flu.

The consequences of official inability to appreciate the gravity of the problem were aggravated by woefully inadequate planning. For example, before the hurricane hit, nobody was in charge of overseeing the response, and nobody had figured out how to avoid conflict over which agency should be in charge of law enforcement so that people could be evacuated effectively. The Louisiana transportation secretary was legally responsible for evacuating thousands of people from hospitals and nursing homes but had no plan in place to do so. The New Orleans Police Department unit responsible for the rescue effort was equipped with three small boats and no food, water, or extra fuel.

For years it was known that 100 000 people lacked transportation out of the city. According to the 2000 census, residents of New Orleans were less likely to own cars than residents of any other city in the United States aside from New York, with its highly developed mass transportation system. Almost all of the New Orleans residents who did not own cars were black, poor, and/or elderly (Berube and Raphael 2005). Yet an evacuation plan for these people was only 10 percent finished when Katrina struck.

Complicating matters further was the inexperience—many say incompetence—of high-ranking officials in the Federal Emergency Management Agency (FEMA). Five of the eight top people in FEMA, including its head, Michael Brown, joined the agency without any experience in disaster management. Many of the agency's top officials, including Brown, were political appointees whose chief claim to fame was loyal service during President Bush's run for the White House. Lacking professional qualifications and relevant job experience, such people often didn't know what to do and didn't appreciate how urgently they had to act. It hardly helped matters that FEMA had lost its independent status after becoming part of the newly formed Department of Homeland Security in 2001 and then suffered budget cuts as resources were diverted to fighting terrorism. Little wonder that the New Orleans relief and rescue effort was slow, inadequate, and often brutal, leading to many deaths (see this chapter's Appendix A).

Tens of thousands of New Orleanians were trapped in the flooded city after the storm passed. Many of them had to be rescued from rooftops by boat or helicopter. Moreover, the bowl was now overflowing with toxic soup—the waters that engulfed the city contained a witch's brew of industrial and household chemicals, sewage, garbage, and rotting human and animal corpses. People were nonetheless forced to wade through the waters to scrounge for drinking water, food, and medicine, and to get to higher ground. Conditions in the Superdome and the New Orleans Convention Center were especially appalling. Thousands of people sought shelter in those buildings, where they were stranded for days in stifling heat, with little or no drinking water, food, medicine, or sanitation. Women miscarried, elderly people died, all suffered horribly.

A Smaller, Whiter New Orleans

Many New Orleans residents returned home after Katrina. Five months after the storm, the city's population stood at two-thirds of its pre-Katrina level (Katz, Fellowes, and Mabanta 2006: 5). Private insurance, individual spending, charitable contributions, and government aid have funded and will continue to fund clean-up, reconstruction, and levee repair and improvement. However, progress has been slow. In 2015, the city's population was one-seventh smaller than it was before Katrina (U.S. Census Bureau 2016c).

Class and race divisions were evident in the storm's immediate aftermath. Some white New Orleans neighbourhoods were extensively damaged by Katrina, and some black districts escaped serious damage. Overall, however, a disproportionate amount of moderate and catastrophic damage occurred in poor, black districts. Thus, if all residents had returned to lightly damaged neighbourhoods and none had returned to moderately and catastrophically damaged districts, the city would have lost 80 percent of its black population and 50 percent of its white population (Logan 2006).

In the city's uptown district, life was starting to return to normal just three weeks after the storm. Along its broad, tree-lined streets, work crews and hired help were restoring the Civil War-era mansions, and well-to-do white families were already living in them. The uptown and other well-off districts (Algiers, the French Quarter, the Central Business District) are on higher ground, and they escaped the worst of the flooding. But buildings in the poor districts, where most of the city's African Americans lived, tended to be on low ground. There, flooding was typically more severe. Three weeks after Katrina, the poor districts were deserted. Much floodwater remained, stagnant and toxic. Emaciated dogs ran wild. Everything was rotting (Mahoney and Freeman 2005). Many of the houses in the poor districts had to be destroyed. Two years post-Katrina, after the Army Corps of Engineers had worked hard and spent a billion dollars to repair the city's hurricane protection system, prosperous neighbourhoods, such as Lakeview, had their flood risk from a big hurricane reduced by nearly 1.7 m (5½ feet). Poor neighbourhoods, such as Gentilly, had their risk reduced by just 15 cm (6 inches) (Schwartz 2007).

It is unclear how many African Americans will eventually return to New Orleans, but an indication of what might lie ahead comes from a survey conducted in Houston between September 10 and 12, 2005, among 680 randomly selected Katrina evacuees. Ninety-eight percent of them came from New Orleans, 93 percent were black, and 86 percent had household incomes of less than US$30 000 a year. Just 43 percent of the respondents said they planned to move back to their hometown ("Survey of Hurricane Katrina Evacuees" 2005).

African Americans are less likely to return than whites are because they lack the money required to do so and, in any case, have less to return to. In addition, evidence suggests that reconstruction efforts have discriminated against the black community. Such efforts first concentrated on less-damaged white neighbourhoods, delaying the reconstruction of predominantly black districts. As time passed and people got settled elsewhere, they were less likely to return. Most government grant assistance went to white, middle-class storm victims. In white districts, government loans to small business were approved at about seven times the rate of loan approval in poverty-stricken black neighbourhoods. Blacks were less likely than whites were to receive insurance settlements that would have allowed them to reconstruct their houses (Bullard 2006).

A 2015 survey found that four out of five white New Orleans residents, compared to just two out of five black residents, think the city has mostly recovered. Since Katrina, the income gap between blacks and whites has widened (Brunhuber 2015). The new New Orleans is smaller and whiter than the pre-Katrina city. As has been the case throughout the history of New Orleans, class and race powerfully shape people's life-chances.

MARKETS, CITIZENSHIP, POWER, AND POLICY

> I think you all know that I've always felt the nine most terrifying words in the English language are, "I'm from the government and I'm here to help."
>
> —*President Ronald Reagan, 1986*

Markets are social relationships that regulate the exchange of goods and services. In a market, the prices of goods and services are established by how plentiful they are (supply) and how much they are wanted (demand). For example, if demand for labour increases and the supply of labour stays the same, the price of labour (the hourly wage) rises. Workers spend and save more, and unemployment falls. In contrast, if the supply of labour increases and demand for labour stays the same, wages fall. Workers earn and spend less, and unemployment grows.

Late-18th-century Britain most closely approximated a completely free market for labour. However, the supply of labour exceeded demand to such a degree that starvation became widespread. The threat of social instability forced the government to establish a system of state-run "poor houses" that provided minimal food and shelter for people without means (Polanyi 1957).

Because perfectly free labour markets periodically cause much suffering and death if left unchecked, they must be regulated by governments. For example, North American laws outlaw child labour, stipulate maximum work hours, make certain holidays compulsory, and specify a minimum wage. Like most people, Canadians and Americans know that without such regulations, the whip of the labour market would destroy many of us.

The rights of people to protection under the law are embodied in the concept of **citizenship**. To varying degrees, citizens of different countries have fought for and won civil rights (free speech, freedom of worship, justice under the law), political rights (freedom to vote and run for office), and social rights (freedom to receive a minimum level of economic security and participate fully in social life) (Marshall 1965). Note the phrase *to varying degrees*. In the fight for citizenship rights, the citizens of some countries have been more successful than the citizens of other countries have been. And note the word *fight*. Citizenship rights are rarely granted because of the grace and generosity of people in positions of power. They are typically extracted by subordinates using force or the threat of force. Consequently, success in achieving citizenship rights depends heavily on how powerful different categories of people are.

Power is the ability to realize one's will, even against the resistance of others (Weber 1946: 180). To oversimplify for the sake of brevity, the balance of power between authorities and subordinates in a given society largely determines how extensive and entrenched citizenship rights become. If subordinates are relatively powerful, citizenship rights become extensive and well entrenched. Laws that ensure broad civil, political, and social rights are passed. Among rich countries, societies like Sweden emerge in the extreme case. If authorities are relatively powerful, citizenship rights do not become extensive and well entrenched. Fewer and weaker laws ensuring civil, political, and social rights are passed. Among rich countries, societies like the United States emerge at the other extreme.

On almost every imaginable measure of citizenship rights, the United States lags behind the other 20 or so other rich countries in the world. For example, long after all adult citizens won the right to vote in other rich countries, many African Americans were still unable to vote. It was only in the 1960s that African Americans won such rights. Nor did the United States lead with respect to women's voting rights. It became the 26th country to grant women the right to vote, following the Scandinavian countries, the British dominions, a number of continental European countries, and the Soviet Union.

The United States compares unfavourably with other rich countries as far as social rights are concerned, too. Thus, the gap between rich and poor is greater in the United States than in any other rich country, and the proportion of the population classified as poor is larger. Americans enjoy no national healthcare system, no national system of paid parental leave, no national system of job retraining, and no national child-care system. In recent decades, the government has slashed the number of families receiving welfare benefits and the cash and non-cash assistance available to each family. As a result, the plight of America's poor—disproportionately composed of children, single mothers with children, and African Americans—has been worsening steadily (Block, Korteweg, and Woodward 2006).

According to Michael Ignatieff (2005), the circumstances surrounding Hurricane Katrina demonstrate that the government of the United States has broken its "contract" with its citizens. I disagree. The contract never stipulated that the American government would care much for its citizens in the first place. True, there have been periods when American governments were charged with greater responsibility. The Great Depression of the 1930s, with its massive nationwide strikes and the civil rights era of the 1950s and 1960s, with its marches, demonstrations, sit-ins, and race riots, were times when Democratic administrations took important steps forward in that regard. But the overall tendency, grown stronger since Ronald Reagan first came to power in 1980, has been for government to minimize its involvement in the lives of its citizens, giving the freest possible rein to the forces of the free market.

I conclude that the tragedy of Katrina was ultimately the result of the imbalance of power between upper and lower classes, and between authorities and subordinates, in the United States.

CRITICAL THINKING EXERCISES

1. Based on this chapter's analysis, list the ways in which natural disasters are influenced by human activity. Do you think that the effects of earthquakes are influenced by human activity or are the effects of earthquakes relatively immune to human influence? Answer by comparing the effects of the October 17, 1989, earthquake in the San Francisco area and the January 12, 2010, earthquake in the Port-au-Prince area. See this chapter's Appendix B.
2. This chapter argues that considerable variation exists from one country to the next in the relationship between state and market. In a page, outline the conditions that might cause the relationship between state and market in the United States to become more like the relationship between state and market in Western Europe. In a page, explain why you think the likelihood of the United States becoming more like Western Europe in this regard is high, medium, or low.

3. A 2001 report by the Federal Emergency Management Agency listed the three most likely disasters to hit the United States: a major terrorist attack on New York City, a severe hurricane hitting New Orleans, and a serious earthquake shaking San Francisco. Based on what you learned in this chapter and what you can learn about the social structure of San Francisco on the Web, write a five-page analysis of the likely effects of a serious earthquake shaking San Francisco.

APPENDIX A

Written Testimony for the Record
by Leah Hodges
Evacuee, New Orleans, Louisiana
Select Bipartisan Committee to Investigate the Preparation for and Response to Hurricane Katrina
U.S. House of Representatives
December 6, 2005

I wish to thank everyone who is listening today for the chance to communicate my story. I come to Congress today representing not just myself, but hundreds, even thousands of other New Orleans residents who experienced the same or similar traumatic experiences and witnessed the same or similar events.

...

Let me begin with a few general points.

1. I don't need to point out the failures of the President, the Governor of Louisiana and the Mayor of New Orleans, as these individuals have already claimed responsibility for everything that happened to us as the result of the hurricane and its aftermath.
2. The people of New Orleans were stranded in a flood and were allowed to die. The military had personnel stationed just 40 miles outside the city, and they could have moved in and got people out sooner. People were allowed to die.
3. Animals from the animal shelter and fish from the fish aquarium were evacuated before the people.
4. The President and local officials issued "shoot to kill orders" and people were shot. People who asked for help were threatened with being shot. My niece and her fiancé, they needed gas. Her fiancé asked military [personnel for] help and they told [him] "if you don't get back inside we will shoot you."
5. Bodies are still being found every day in New Orleans. Most people in New Orleans do not believe the official body counts.
6. The devastation that hit New Orleans was foreseeable and avoidable, and because it was not avoided, New Orleans was turned into a mass grave.
7. As a hurricane survivor, I and my family were detained, not rescued.

My family was ordered to evacuate our home. We were directed to evacuation points. Beforehand, I, my mother, my brother and two sisters visited a nursing home where the elderly clients had been abandoned by the owners and staff. There were five elderly persons there; the others had been evacuated earlier, perhaps by family. The day before the flood, the manager had come and told everyone they had to get out. Taking the keys to the bus that the home used to transport the senior citizens, the manager left them stranded. We rescued them. We shared all our food and provisions. When we approached the police and asked for help, the[y] refused to help us. Instead, they threatened to shoot my baby brother.

We were then lured to the so-called evacuation points. This was several days after the hurricane had struck. The city was flooded. Soldiers had showed up with M16s and military weapons. They had declared New Orleans and Jefferson Parish a war zone. They loaded us onto military trucks after they told us they would take us to shelters where our basic needs would be met.

We were dropped off at a site where we were fenced in, and penned in with military vehicles. The armed military personnel brought in dogs. There we were subjected to conditions only comparable to a concentration camp.

We were in a wide open space along the interstate and under the Highway 10 causeway. The overpass provided little shade, however. During the days, we were exposed to the hot sun. August is the hottest month in New Orleans. It was early September and still extremely hot. Our skin blistered. My mother's skin is still not fully healed.

We were just three miles from an airport, but we were detained there for several days. Many of those who were there when we arrived had already been there several days. On any given day there were at least ten thousand people in the camp. On my last day there, I would estimate there were still three thousand detainees. By that time, nearly all the white people had been selected to evacuate first. They were put on buses and shipped out, leaving the remaining population 95 percent black.

There was muck and trash all over the causeway. Nothing was done to clean it up. At night, we were subject to sleep deprivation as low-flying helicopters were deliberately flown right over us. They would throw up the muck and trash, so that it would get all over us, even the pregnant women, the elderly, the infirm.

The military did not bring anything to help keep any of us alive. Not even a first aid kit. But they had body bags. They were doing nothing for the pregnant women. Some women miscarried. I know that conditions at the Convention Center were much the same. My niece was there. She was pregnant and she was terrified that her unborn baby had died.

When she asked the military for help, they told her to wait until she was sure the baby was dead and then talk to them.

When I later spoke of my experience to a state trooper, he told me: "I would have rebelled." They set us up so that we would rebel, so that they could shoot us. At one point they brought in two truckloads of dogs and let the dogs out.

We would circulate through the camp to assist the sick and elderly and pregnant. One day, when I was on my way to get some water, I met a friend. He was a fellow musician. He told me that he wanted to try to get word out to the news media. But he was afraid to leave his family. I told him I would look after his family. But while he was gone I also had to circle back and check on my own family. I found that my brother had come up with an idea. He had persuaded a woman who was pregnant and due for labor to fake as if she were in labor. They told those in charge that she needed medical attention or she could have a miscarriage, and that got her out.

There was an old man from the senior center, he was an amputee. We had to carry him to the bathroom. They would not assist in caring for our people. The heat was unbearable. We got to the point we were so afraid of losing him to a heat stroke. We told them he was in a diabetic coma, that's how we got him out.

Mother is a cardiac patient, born with an enlarged heart. She suffers extreme hypertension. For three days I pleaded with them for care, and they would not do anything. Finally, on the third day, someone came out to check her blood pressure. The sphygmomanometer did not appear to be in working condition. I told the man, who was from the Coast Guard, to take my blood pressure first. The thing fell to pieces in his hands. It never worked.

The camp was so big, and people were scattered. People were deliberately kept apart. One woman was not allowed to see her two children.

At the camp, they lied and told us all the buses were going to the same place. They wouldn't tell us when the buses were coming. Meanwhile, my Mother sat in the blazing hot sun ...

On the last day they refused to give food and water to the ill for 24 hours.

People died in the camp. We saw the bodies lying there.

They were all about detention, as if it were Iraq, like we were foreigners and they were fighting a war. They implemented war-like conditions. They treated us worse than prisoners of war. Even prisoners of war have rights under the Geneva Convention.

Source: U.S. House of Representatives (2005).

APPENDIX B

Nothing Learned: The Great Port-Au-Prince Earthquake of 2010

The number of people killed in an earthquake depends on many factors, including the amount of energy released by the quake, its depth, and the population density at the earthquake epicentre. No two earthquakes are identical, even if, superficially, they appear almost the same. For example, the earthquake that struck near Port-au-Prince, Haiti, in 2010 registered 7.0 on the Richter scale, while the earthquake that hit the San Francisco Bay area in 1989 registered 6.9. These numbers may make the quakes, or temblors, seem nearly identical. However, the Haiti earthquake released about 40 percent more energy and occurred 5 km (3 miles) closer to the earth's surface. On the other hand, about 6 million people lived in the San Francisco Bay area compared to about 3 million in the Port-au-Prince region. The death tolls? In San Francisco: 63. In Port-au-Prince: about 220 000.

The discrepant death tolls were less the result of differences in the magnitude of the disasters than in the preparedness of San Francisco and Port-au-Prince to deal with them. A modern San Francisco structure such as the famous pyramid-shaped Transamerica Building is earthquake-proof because its base rests on giant pillars that are isolated from the surrounding earth, and its walls are reinforced with steel cross-braces. The Transamerica Building swayed about 30 cm (about a foot) in 1989 but suffered no damage. In contrast, the typical structure in Port-au-Prince is made from hollow cinder blocks held together by substandard mortar. Such a structure can hold quite a lot of weight, but even slight side-to-side movement will cause it to collapse. If rebar (a steel reinforcing rod in concrete) is used in larger buildings, and it likely isn't, it is half the diameter of the steel reinforcing rods used in modern buildings in industrialized countries. One of the few buildings in Port-au-Prince that completely withstood the 2010 earthquake was the American embassy, constructed according to U.S. building code (Perlman 2010).

Haiti's infrastructure was so weak and inadequate because it was by far the poorest country in the western hemisphere. Yet it was not always so (Henley 2010). Two centuries earlier, Haiti was producing great wealth. Abundant sugar, coffee, and indigo plantations made it France's richest colony in the New World. Sixty percent of Europe's coffee and

40 percent of its sugar came from Haiti. Not that the hundreds of thousands of slaves who were brought over from Africa to create that wealth saw any of it. They laboured and lived in the most wretched conditions imaginable; their life expectancy was a mere 21 years. The wealth flowed back to the French treasury.

In the early 19th century, a bloody slave revolt led to emancipation and independence, but in exchange for recognizing the new country, France demanded—and received—tens of millions of francs in gold between 1825 and 1947. To find the money for these reparations, the government of Haiti borrowed heavily from American, German, and French banks. By 1900, fully 80 percent of Haiti's national budget was going to loan repayments, so almost nothing was left for the welfare of its citizens.

A second revolution broke out in 1911. The United States became so terrified that Haiti would default on its debt that it occupied the country for 20 years. During the Great Depression (1929–39), the country was unable to export much of anything, and with the United States, France, Britain, and Germany supporting rival political factions, political disunity persisted and economic decline continued.

Haiti fell under the unfeeling dictatorships of François ("Papa Doc") and Jean-Claude ("Baby Doc") Duvalier from 1957 to 1986. While having dissenters and rivals tortured and killed, father and son embezzled much of the aid Haiti received from other countries and enjoyed the benefits of huge loans they took out in the name of their people. When Baby Doc fled the country in 1986, he took an estimated $900 million with him. He left a government broken and corrupt; an economy lacking infrastructure, know-how, and investment; and a people miserable and impoverished.

And so Haiti remains today.

Three years after the 2010 earthquake, 350 000 Haitians still lived in tents while uncounted others had returned to substandard housing. Cholera had killed more than 7500 people. Food shortages remained common. A poorly trained and organized Haitian government bureaucracy was able to help the situation little. Billions of dollars in reconstruction aid had been promised by foreign donors, but much of it never materialized. Most of the money that did arrive was spent on temporary relief and the salaries of foreign staff ("Haiti Still Waiting" 2013).

Was this hell the handiwork of nature or of people?

PREVENTION
CURE

5

The Social Bases of Cancer

With the assistance of Carey Bennett, Yolen Bollo-Kamara, Christina Coliviras, Joanne Courneyea, Karishma Hossain, Leah Hui, Mahdi Hussein, Nhi Huynh, Abinaya Krishnamoorthy, Sophia Li, Meirui Li, Katelyn Li, Randeep Nijjar, David Pham, Nicole Rodrigues, Joobin Sattar, Ioana Sendroiu, Jessica Syrette, Hoyee Wan, Ted Wu, Anoosha Zafar, Khuzaima Zafar, and Yunshu Zhao.[1]

THE SCOPE OF THE PROBLEM

I haven't had a cigarette for more than three decades. I tried my first smoke with Brian Robertson behind the garage in 1965 when I was 14. It seemed cool. Besides, it was fun doing something adults told us not to do. Within three years, I was up to a pack a day.

At 18, I failed to implement my New Year's resolution to stop smoking. I needed a powerful motivator to quit, and it didn't come along until the birth of my first daughter 14 years later. I vowed not to surround her with dangerous smoke and risk having her pick up my bad habit.

[1] In spring 2010, I announced in my SOC101 classes at the University of Toronto that students with an A average could apply to help me review literature and analyze data for this chapter during the summer. I had about 2300 students, approximately 350 with an A average. About 90 applied. I selected about two dozen students whom I regarded as best suited for the job. Seventeen of them were entering their second year of studies, and the rest, their third or fourth year. The group consisted of 12 social science majors and 11 life science majors. They were all volunteers who agreed to work on the project at least five hours a week for three months without pay. They worked enthusiastically and taught me a lot. I am deeply grateful to them for their assistance and for giving me the opportunity to get to know them better.

Quitting was hell. I still occasionally crave a smoke. However, I can never light up again because I am an addict, and I know that smoking will make me feel unhealthy and put others at risk while likely sending me to an early grave.

With variation in detail, millions of people have lived my story. The association between smoking and lung cancer was first officially proclaimed a year before I smoked my first cigarette (U.S. Department of Health, Education, and Welfare 1964). Governments and public health officials immediately launched a massive educational campaign, and it helped cut substantially the percentage of North Americans who light up. In 1965, 50 percent of Canadians over the age of 14 smoked daily or occasionally. In 2011, the figure stood at 20 percent (27 percent among Canadians between the ages of 20 and 34) (Physicians for a Smoke-Free Canada 2012). Still, more than a quarter of all deaths among Canadians over the age of 35 are due to tobacco use, and tobacco kills half of all lifetime users (Remennick 1998: 17; World Health Organization 2002: 36–37). If you smoke, you might be lucky and avoid a wretched and premature death, but I wouldn't count on it.

In 1971, U.S. President Richard Nixon declared a "War on Cancer." His campaign mobilized educators, researchers, physicians, and public health officials in North America to devote more energy and resources to fighting the disease. Four decades and more than US$100 billion later, how have we fared?

To assess the impact of interventions, we cannot simply compare the number of cancer cases at two time points, in part because the Canadian population has grown and there are bound to be more cancer cases among more people, all else the same. To remove the effect of population growth from our comparison, we must examine the **cancer incidence rate**, that is, the number of cases per 100 000 people. Doing so enables us to compare the number of cancer cases among the same number of people, even though the population as a whole has increased.

We face a second problem with cross-time comparisons. If the population is aging over time, as it is in Canada, we would expect the cancer incidence rate to increase, all else the same. Older people are more likely to develop cancer than younger people are, so we must also remove the effect of population aging from our comparison by calculating the **age-standardized incidence rate** of cancer at different time points. We do this by manipulating the data so that the age distribution of the population at the later time point is the same as at the earlier time point. Doing so ensures that any change we observe in incidence rates is not due to population aging (Wells 2010).

The four most common cancer sites in Canada are the prostate, the lungs, the colon and rectum, and the breasts. The age-standardized incidence rate of prostate cancer increased 65 percent between 1973 and 2010 (see Table 5.1). The age-standardized lung cancer incidence rate fell by 13 percent for men but rose 269 percent for women because more and more women began smoking in the 1960s and 1970s. The age-standardized incidence rate of colorectal cancer fell 13 percent for women but increased 13 percent for men between 1973 and 2010. The age-standardized incidence rate for breast cancer rose 22 percent. Between 2001 and 2010, the overall age-standardized incidence rate for all types of cancer rose by 0.5 percent per year for females and decreased by 0.7 percent per year for males.[2] We declared a war on cancer and, on some fronts, cancer is winning.

On other fronts we have enjoyed the upper hand. The **mortality rate** from cancer (the number of deaths per 100 000 people) has been declining slowly as medical researchers develop new tests for early detection and improved ways of keeping people with cancer alive. The age-standardized lung cancer incidence rate for men has been falling since 1984, and the rate for women has been falling since 1998—striking evidence of what can happen when many people remove cancer-causing agents (in this case, tobacco) from their lives. The horrifying fact remains, however, that cancer is the leading cause of death in Canada. In this country, about 42 percent of women and 45 percent of men develop cancer, and approximately a quarter of Canadians die of the disease (Canadian Cancer Society Advisory Committee on Cancer Statistics 2015: 6, 16).

The Genetic Revolution

The year 2000 sparked new hope in the war on cancer. Scientists completed the first draft of the human genome—the "map of life" encoded in our DNA—at a cost of US$3 billion. With the draft in hand, Dr. Francis S. Collins, head of the Human Genome Project, proclaimed

[2] The introduction of early screening accounts for part of the increased incidence of some forms of the disease. Thus, in the mid-1980s, the incidence rates of prostate and breast cancers spiked following the increased use of blood tests to detect the prostate-specific antigen and mammograms to detect breast tumours. However, early detection also increases the risk of misdiagnosis and mistreatment (see, for example, Saul 2010).

Table 5.1 Age-Standardized Cancer Incidence Rates (per 100 000 People) for the Four Leading Cancer Sites, Canada, 1973 and 2015

	Cases per 100 000 Men		**% Change for Men**	**Cases per 100 000 Women**		**% Change for Women**
Cancer Site	**1973**	**2015**	**1973–2015**	**1973**	**2015**	**1973–2015**
Prostate	60	99	65	n.a.	n.a.	n.a.
Lung	67	58	–13	13	48	269
Colon and rectum	53	60	13	46	40	–13
Breast	n.d.	n.d.	n.d.	82	100	22

Notes: n.a. = not applicable; n.d. = no data (less than 1 percent of all breast cancer occurs in men, making the incidence rate too small to include.)

Sources: Canadian Cancer Society (2002: 42, 44); Canadian Cancer Society Advisory Committee on Cancer Statistics (2015: 2).

that within a decade, genetic tests would be available for the 25 leading causes of illness and death, allowing doctors to identify and treat people with predispositions to particular diseases even before they showed signs of illness (Altman 2000).

To carry out Dr. Collins's plan, scientists conducted nearly 400 studies costing several million dollars apiece over the next decade. They focused on common genetic variants, the roughly 0.5 percent of DNA that can differ between any two individuals. They hoped find associations between specific genetic variants and particular diseases, and then use that information to predict disease susceptibility in individuals.

They found about 850 statistically significant associations but also discovered that, if a person had a genetic variant associated with a type of cancer or some other disease, it was usually a poor predictor of whether the person would get sick. Many people with the variant never developed cancer, and many people without the variant did. Researchers had to go back to square one. Although some scientists now believe that refocusing research on rare rather than common genetic variants will eventually help them understand the genetic roots of disease, others believe that, in most cases, these roots are so complex they will not be susceptible to drug treatment (Cancer Genome Atlas Network 2008; Jones et al. 2008; Parsons et al. 2008; Robertoux and Carlier 2007; Wade 2010).

However, even if genetic research fulfills Dr. Collins's promise, genes *by themselves* are responsible for just 5 to 10 percent of all cancers and a substantially smaller percentage of cancers in people over the age of 25 (Baird 1994). In other words, although genetic mutations cause all cancers, environmental factors cause more than 90 percent of the genetic mutations leading to cancer (Fearon 1997; Hoover 2000; Kevles 1999; Lichtenstein et al. 2000; Wu et al. 2016). *Environment* in this context means "anything that people interact with, including exposures resulting from ... what we eat, drink, or smoke; natural and medical radiation ...; drugs; socioeconomic factors that affect exposures and susceptibility; and substances in air, water, and soil" (U.S. Department of Health and Human Services 2009: 1.)

The Cancer Paradox

Here we arrive at the paradox that motivates this chapter. If environmental carcinogens cause more than 90 percent of the genetic mutations resulting in cancer, it follows that we can win the war on cancer by eliminating the offending substances or at least drastically reducing their prevalence and our contact with them. Unfortunately, the structure of our society makes it difficult for us to do so. Many people lack the resources that would allow them to take preventive measures. Moreover, industry and government resist change. This situation creates the **cancer paradox**: Although we know how to reduce the cancer incidence rate drastically, we have not been able to move vigorously on this front.

Although the estimated percentage of cancer deaths caused by tobacco and diet given in Table 5.2 are undoubtedly high, these two sources account for most deaths from cancer.[3] To lower the death rate from these sources, people need to stop using tobacco and change their diets to adhere to scientific recommendations.

In 2007, a panel of 22 leading experts from five countries completed a five-year review of all research relating to diet and cancer. Table 5.3 shows the foods and physical conditions they judged to be "convincingly associated" (relatively strongly correlated at a high level of statistical significance) with increased risk for various cancers. Among other things, they found that being overweight, eating red and processed meat frequently, and drinking excessive amounts of alcohol significantly increase the risk of getting cancer. Based on their findings, they made eight general recommendations:

1. Stay as lean as possible within the normal range of body weight.
2. Be at least moderately physically active, equivalent to brisk walking for at least half an hour every day.
3. Avoid high-calorie foods, such as sugary drinks and fast food.
4. Eat mainly foods of non-starchy plant origin.
5. Limit consumption of red meat to one portion a week and avoid processed meat.
6. Limit alcohol consumption to two drinks a day for men and one drink a day for women.
7. Avoid salty foods, and do not eat mouldy grains and legumes.
8. Aim to meet nutritional needs through diet, avoiding dietary supplements.

[3] Table 5.2 excludes older adults and black people (who are more likely than younger people and white people to get cancer) and post-1977 data (which is problematic because recent decades have witnessed exponential growth in the use of chemicals in industry and in industrial and household products).

Table 5.2 Cancer Deaths and Their Proximate Causes, White Americans Younger Than 65, 1940s to1970s

Causes	Approximate % of Cancer Deaths
Diet	35
Tobacco	28
Infection[a]	9
Purely genetic factors	8
Reproductive and sexual behaviour[b]	6
Occupation	5
Natural radiation[c]	3
Pollution	2
Medicines and medical procedures[d]	1
Food additives	1
Industrial products	1
Total	99[e]

Notes:

[a] Especially stomach cancer caused by the *H. pylori* bacterium, cervical cancer caused by the human papilloma virus, and liver cancer caused by the hepatitis B and C viruses.

[b] Especially breast cancer because of late childbirth and low fertility, and cervical cancer because of poor personal and sexual hygiene and promiscuity (the latter of which increases the likelihood of contact with the human papilloma virus).

[c] Especially skin cancer caused by excessive exposure to sunlight and lung cancer caused by excessive exposure to radon.

[d] Including radiation from medical imaging.

[e] Does not equal 100 because the figures are only approximate.

Sources: Adapted from Doll and Peto (1981: 1256); Remennick (1998: 17, 32).

It is significant that people who know about these recommendations and act on them tend to be comparatively well-educated and well-to-do. They are also relatively likely to have regular physical examinations and screening procedures that may allow doctors to detect cancer early and deal with it before it becomes life-threatening. People who do not know about these recommendations or know but fail to act on them tend to be less well-educated and well-to-do. They are more likely to smoke, drink excessive amounts of alcohol, lack daily exercise, and eat foods high in animal fat, sugar, salt, and starch. Not surprisingly, therefore,

Table 5.3 Foods and Physical Conditions "Convincingly Associated" with Elevated Cancer Risk

Cancer Site Foods and Conditions	Colon, Rectum	Lung	Breast (women)	Mouth, Pharynx, Larynx	Esophagus	Liver	Pancreas	Endometrium (women)	Stomach	Kidney
Body fatness	X		X		X		X	X		X
Body height	X		X							
Red meat	X									
Processed meat	X									
Alcoholic drinks	X[a]		X	X	X					
Salt									X	
Arsenic in drinking water		X								
Beta-carotene supplements[b]		X								
Aflotoxins[c]						X				

Notes:

[a] For men. There is "probable increased risk" associated with alcoholic drinks for women.
[b] For smokers. Beta-carotene is an inactive form of vitamin A.
[c] Naturally occurring in fungus found in mouldy cereals, spices, and nuts.

Source: Reprinted with permission from the American Institute for Cancer Research.

the lower a person's education and income, the more likely that person is to get cancer (Cancer Care Ontario 2010; Glazier et al. 2004; Johnson et al. 2010; Link and Phelan 1995; Mao et al. 2001; Phelan et al. 2004; Pruitt et al. 2009; Siahpush et al. 2006).[4]

Even when people with low income and less education know about the need to stop smoking and improve their diets, urging them to do so tends to have little effect. That is because the conditions of their existence conspire against them. At work, they are more likely to experience little control and high stress than are people with higher income and more education. At home, they are more likely to face marital stress and divorce because of money problems. They are, therefore, more likely to turn to tobacco and alcohol to help them feel better in the short term.

They are also more likely than are people with more education and higher income to be exposed to carcinogens at work. Miners, construction workers, people who work with asbestos, welders, petroleum refinery workers, rubber industry workers, textile industry workers, footwear production and repair workers, hairdressers, farmers, paint manufacturing workers, house painters, furniture and cabinet makers, machine shop workers, and garage mechanics all have elevated cancer incidence rates. In their neighbourhoods, they tend to suffer relatively high exposure to air and water pollutants that contribute to cancer risk because, on average, they live closer to dirty industries and industrial waste sites than others do. Because of their low income, they often choose unhealthy foods because these are within their budget while healthy foods are too expensive (Brophy et al. 2006, 2007; Keith and Brophy 2004; Remennick 1998: 25–38). The notion that people in such circumstances are free to lower their cancer risk is naive, to say the least.

Given the situation just described, you might think that governments and voluntary associations concerned with cancer would be investing heavily in research on how disadvantaged people can gain the educational, organizational, and political resources needed to change health, environmental, industrial, and social policy in a way that would lower their exposure to cancer risk. However, if we break down the $402 million that the Canadian government and voluntary associations spent in 2007

[4] Some studies of the relationship between class and cancer use occupation as an indicator of class position, thus excluding unemployed workers and people who have left the job market because they cannot find work. Such people (today, about 15 percent of the Canadian adult population) are especially prone to cancer (Remennick 1998: 30). If they were included in the studies, the observed correlation between class and cancer would be even stronger. Note also that breast cancer is an exception to the pattern just described. It is especially common among women who delay childbirth and have relatively few children; such women tend to have higher incomes and more education.

on cancer research, we find nothing of the kind. Some 55.1 percent of this sum went to investigating the biology and causes of the disease. Another 43.1 percent went to research on detection, diagnosis, prognosis, treatment, and related efforts. Just 1.8 percent— $7 million—was allocated for prevention (Canadian Cancer Research Alliance 2009: 22, 28). These sums refer just to research budgets. If we added the medical costs of dealing with cancer, the imbalance between the massive effort aimed at cure and the paltry effort aimed at prevention would be even more glaring.

If funding priorities remain what they are, it is because powerful industrial interests, governments, and segments of the population benefit economically from a bias toward finding cures rather than taking preventive action. A reorientation of policy would cause them financial harm. Even part of the medical establishment may fear that its research and treatment budget would shrink if we gave prevention higher priority than it presently has.

If you need evidence of the cancer paradox, you might visit the area around Fort McMurray, in the middle of the Alberta oil sands.

THE ALBERTA OIL SANDS

Crude bitumen is a naturally occurring form of petroleum. About 175 billion barrels of the substance cover an area of central and northern Alberta that is bigger than England (see Figure 5.1).

Only Saudi Arabia's and Venezuela's petroleum reserves are bigger than Alberta's (Alberta Geological Survey 2009; U.S. Energy Information Administration 2009). However, Alberta crude bitumen is considerably more difficult to extract than Venezuelan and especially Saudi oil is. That is partly because it pours about as well as cold molasses. To make matters worse, it is mixed with sand and clay. If the mixture lies near the surface, giant cranes dig it up and massive trucks haul it to upgraders that remove the waste material. If it is far below ground, miners inject steam so the oil sands can be pumped to the surface for upgrading.

Upgrading is costly, dirty, and energy- and water-intensive. The first step involves separating out the crude bitumen by mixing oil sand with hot water and caustic soda. This procedure creates a toxic waste sludge that is pumped into artificial lakes known as "tailing ponds." Even if current government plans to limit the size of tailing ponds are enforced, enough sludge will be produced by 2020 to form a cube more than 1.1 km (almost three-quarters of a mile) on a side (Simieritsch, Obad, and Dyer 2009). If that amount of sludge were dumped into a basin the

Figure 5.1 The Alberta Oil Sands

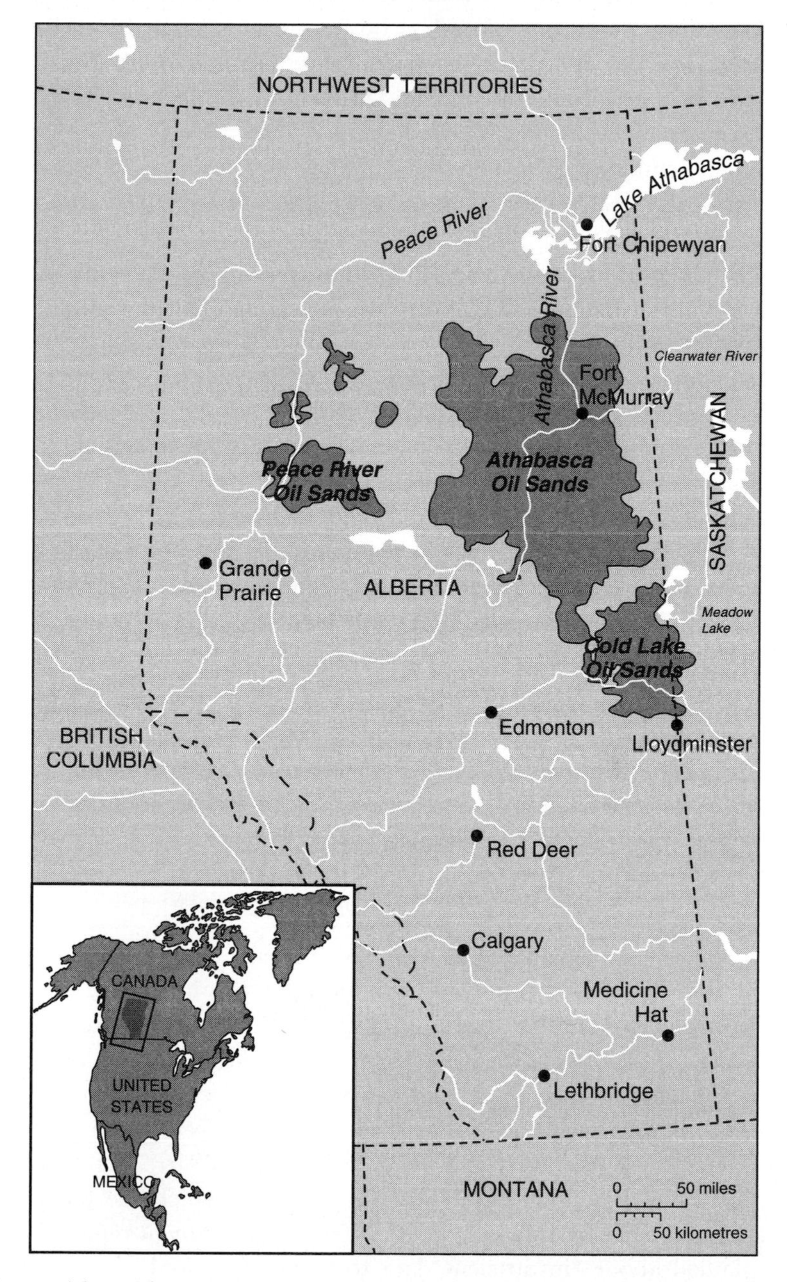

Source: Adapted from Nikiforuk (2008: 10).

size of Lake Ontario, the world's 11th biggest lake at 19 259 square km (some 7500 square miles), it would be 56 cm (almost two feet) deep.

The tailing ponds are located on both sides of the Athabasca River. Shifting foundations, collapsing walls, seepage, and other problems allow sludge to pour into the river and leech into groundwater. Although the Alberta government has said it will take action to limit the problem, seepage is estimated to be more than 25 million litres (some 5.5 million gallons) a day (Energy Resources Conservation Board 2009; Timoney and Lee 2009: 72).

The sludge is highly toxic, containing dozens of known cancer-causing chemicals, including benzene, arsenic, and 25 varieties of polycyclic aromatic hydrocarbons (PAHs). Carcinogens are also emitted into the water and the air as crude bitumen is converted into synthetic crude oil and then refined. PAH levels in the sediment of the Athabasca River are about twice as high as the level observed to induce liver cancer in fish. Arsenic levels are about a third higher than the guideline for protection of aquatic life. Scientific studies suggest that the number of fish deformities is higher than expected, increasingly common, and related to water quality (Timoney 2007; Timoney and Lee 2009). Here is what one long-time local fisher had to say about the problem:

> There's deformed pickerel in Lake Athabasca.... Pushed in faces, bulging eyes, humped backs, crooked tails ... never used to see that. Great big lumps on them ... you poke that, it sprays water.... A friend caught a jackfish recently with two lower jaws... He had seen deformed jackfish before, but never one with two jaws.... The skins on the whitefish are starting to turn red. Before they used to be white.... What's in the water? ... One of the healthiest lakes in Canada is now one of the deadliest lakes. (Ray Ladouceur in Timoney 2007: 62)

Evidence is mounting that substantial concentrations of carcinogens are in the air, surface water and groundwater throughout the region (Kelly et al. 2009, 2010; Liggio et al. 2016; Miall 2013; Percy 2013; Schindler 2013). Scientists have identified hundreds of regional sites where substances "pose or are likely to pose an immediate or long-term hazard to human health or the environment" or exceed legally permitted levels (Treasury Board of Canada Secretariat 2016). Yet remarkably little has been done to assess the extent to which the oil sands pose a human cancer threat; a series of recent government reports have complained about "insufficient data to monitor ... and "assess [the] impact of the oil sands" (Wallace 2013: 168).

Many people are scared. For example, a few years ago, a doctor in Fort Chipewyan, a largely First Nations community of about 1100 people on the lower Athabasca River, 250 km (155 miles) downstream from the Alberta oil sands, observed elevated levels of cancer in the local population. Residents drink untreated water when they go hunting. Many eat local fish, vegetables, berries, moose, deer, and rabbit, all of which are exposed to carcinogens in the water and the air. The cancer incidence rate in Fort Chipewyan was 31 percent higher than expected between 1995 and 2006, taking into account the age structure of the population and the proportion of First Nations people in town. Blood, bile duct, lymphatic, and soft tissue cancer were responsible for the higher-than-expected cancer incidence rate (Chen 2009).[5]

Industry remains a cheerleader of rapid, market-driven oil sands development. In the face of criticism, the Canadian Association of Petroleum Producers (CAPP) launched a multimillion-dollar public relations campaign, calling it a "celebration of Alberta and the energy industry" (Collyer 2010: 13). Historically, the Alberta government has not been far behind in its support of the industry. As a former premier once said, "there's no such thing as touching the brake" on oil sands development (quoted in "Stelmach Prepares" 2006).

Such enthusiasm should not be surprising. The oil and gas industry is responsible for the creation of half a million jobs in Alberta and elsewhere in Canada. Fully a third of the Alberta government's revenue comes from the oil and gas industry (Collyer 2010: 4–5).

Yet most Albertans are cautious. One poll[6] found the following:

- Seventy percent of Albertans disagree that oil sands development should proceed as quickly as possible.
- Seventy-one percent agree that new oil sands projects should not be approved until infrastructure and environment management issues are addressed in the oil sands region.
- Seventy-four percent want the government to manage oil sands development rather than allowing market forces to determine the pace of development.
- Eighty-three percent want more government investment in environmental protection in the oil sands region.

[5] We must treat the findings cautiously because the population of Fort Chipewyan is so small Just 12 "extra" cancer cases account for the study's main finding. However, the results are of sufficient concern that the government planned a full-scale community health assessment (Svenson 2010).
[6] Pembina Institute (2007). Reprinted with permission.

It thus seems that most Albertans understand that slower, more environmentally sensitive development is prudent. Most other Canadians seem to be of the same mind, as can be seen in ongoing debates concerning pipeline construction.

THE GREAT CANADIAN PIPELINE DEBATE

Alberta supplies most of the oil Canadians need. Most of the oil it can't sell in Canada is shipped to the United States. However, north–south pipelines run mainly to refineries in the midwestern United States, where an oil glut exists. This situation depresses the price that Alberta oil can fetch in the U.S. market, costing oil companies and Canadian taxpayers billions of dollars a year in lost oil revenue and forgone royalties. The loss is expected to become more acute as oil sands production rises.

Three new pipeline routes have been proposed to relieve the glut—one from Alberta to the British Columbia coast, one extending existing pipelines from Ontario through Quebec and New Brunswick, and one extending existing pipelines from Alberta to Nebraska, from which the oil can then be piped to refineries in Louisiana. From the Pacific, Atlantic, and Gulf coasts, most of the oil will be shipped to Asia.

The benefits of the proposed projects are substantial. They will increase corporate revenue and government royalties, end eastern Canada's dependence on foreign oil, and create construction and refining jobs. However, environmental concerns are also serious. For example, many Quebeckers fear transporting oil by any means after the 2013 Lac-Mégantic rail disaster that killed 47 people.

Americans are also worried about oil spills. The original plan routed the southern pipeline across one of the continent's largest underground water reserves. Modification of the route in response to protests largely took care of that problem, but a second objection remains: Oil sands production is especially tough on the environment. It requires much more energy and water than conventional oil production does, and it causes more pollution, including the dumping of carcinogens into the air and water systems, especially in northern Alberta, where many First Nations people reside. Most First Nations people in British Columbia note that the proposed western pipeline will pass across First Nations territory, including river systems. They worry about damage to their land due to construction and inevitable oil spills. Many regard the proposal as an example of **environmental racism**, the tendency to heap environmental dangers on people who are disadvantaged, and especially on members of disadvantaged racial minority groups.

OTHER CANCER HOT SPOTS

There are many other cancer hot spots in Canada. Consider Windsor, Ontario, a medium-sized city with a metro population of about 320 000. It is home to a minivan assembly plant, two engine plants, tool and die manufacturers, scrap metal recycling plants, and a foundry. It is downwind of the Detroit area's steel mills, chrome-plating plants, municipal waste incinerator, coal-burning power plant (converted to natural gas in the late 1990s), and wastewater treatment plant and associated sludge incineration facilities.

Apart from this heavy industry, more than 10 000 diesel-burning trucks cross the Ambassador Bridge connecting Windsor and Detroit every weekday. Ongoing air quality assessments find a long list of carcinogens that exceed accepted risk levels: chromium, benzene, formaldehyde, PAHs, cadmium, and so on. Not surprisingly, Health Canada data show that Windsor has a cancer incidence rate significantly higher than the rest of Ontario's. For example, in the 45 to 74 age cohort, cancer incidence is 10 percent higher for men and 5 percent higher for women (Gilbertson and Brophy 2001). Similar stories could be told about Sarnia (Ontario), Sydney (Nova Scotia), and other Canadian towns and cities with heavy industry (Barlow and May 2000; Keith and Brophy 2004; *Toxic Trespass* 2006).

Often, government and especially industry are reluctant to publicize elevated levels of carcinogens in the air and water, let alone act on them (Egilman and Howe 2007). For example, Health Canada would not initially release the Windsor data because environmental agencies were concerned about the clean-up cost, medical officers of health worried about the danger of disseminating "uninterpreted" data to the public, and government fretted about its liability and the cost of remedial action. The jig was up only when a CBC reporter produced a copy of the data during a filmed interview with a Health Canada official (Gilbertson and Brophy 2001: 828). Even then, government did little to remedy the situation.

Similarly, official foot-dragging also meant that 30 years passed between the detection of elevated levels of toxins in Nova Scotia's Sydney Harbour and in local lobsters (1980) and the beginning of a serious clean-up effort (2010), completed only in 2013. Until then, people crowned the Sydney tar ponds, repository of a century of sludge from the coke ovens of the city's steel mill, Canada's most toxic waste site. The sludge oozed into the gardens and basements of people living in the neighbourhoods bordering the ponds (Barlow and May 2000). Cape Breton County, where Sydney is located, had the highest cancer incidence rate in Nova Scotia, and Nova Scotia became the province with the highest cancer incidence rate in Canada.[7]

Other Canadian examples abound. At latest count, the federal government had identified 5339 toxic waste sites and another 2353 suspected toxic waste sites in the country and, as you will see, the list is far from complete (Treasury Board of Canada Secretariat 2016).

CORRELATION, CAUSE, AND COSMETICS

Ontarians born under the astrological sign of Sagittarius are 28 percent more likely than other Ontarians to land in the hospital because of a broken arm ("Why So Much" 2007). How can we explain this fact? An astrologist might say that the typical Sagittarian is energetic, adventurous, and inclined to risk-taking and therefore prone to arm-breaking. A scientist is more likely to conclude that Sagittarians' arm-breaking tendency is no more than a statistical fluke. After all, if you sample enough people and look hard enough, you will find all kinds of statistically significant correlations. However, only some of them will meet the four strict criteria for establishing the existence of a **cause-and-effect relationship** on scientific grounds: (1) the cause must precede the effect; (2) cause and effect must be significantly correlated;[8] (3) an observable mechanism must link the cause with the effect; and (4) background factors that could plausibly be responsible for the correlation between the cause and effect must be

[7] Data for the period 2000 to 2004 show the overall cancer incidence rate in Cape Breton County dropping to the provincial average, although lung and colorectal cancer incidence rates remained significantly above the provincial average (Cancer Care Nova Scotia 2006: 8–13).

[8] Epidemiologists traditionally break criterion 2 into (2a) higher exposure must lead to more acute or faster outcomes, and (2b) the correlation must hold under different conditions, at different times, and in different places (Hill 1965).

ruled out. Explanations for why things happen often falter on points 3 and 4. For instance, from a scientific point of view, the lack of an observable mechanism linking presumed cause (being born under the sign of Sagittarius) and presumed effect (landing in hospital with a broken arm) renders the astrologer's explanation dubious.

Similarly, early efforts to establish that smoking causes cancer met the first two criteria for establishing a cause-and-effect relationship but foundered on criteria 3 and 4. Scientists could not immediately identify the biological mechanisms that caused cells to mutate and become cancerous, nor could they rule out that background factors associated with smokers, such as eating unhealthy food and living a sedentary lifestyle, accounted for the tendency of smokers to get cancer. It took decades of research to meet criterion 3, in particular. In the interim, tobacco companies could deny the existence of a proven cause-and-effect relationship between smoking and cancer.

It is significant, however, that public health officials and governments did not wait until they had strict causal proof before launching a massive educational campaign aimed at getting people to stop smoking. They observed such a high correlation between smoking and cancer, and had so many convincing animal studies, case reports, toxicology studies, and other forms of evidence in hand that they decided to act quickly to prevent hundreds of thousands of premature deaths in North America each year. Obviously, they were right to do so.

We find ourselves in a similar predicament today with many suspected carcinogens. The U.S. Department of Health and Human Services lists 54 known human carcinogens and 183 substances "reasonably anticipated" to be human carcinogens to which people are significantly exposed (U.S. Department of Health and Human Services 2009; see also CAREX Canada 2010). They include pesticides, flame retardants, phthalates (found in cosmetics and shampoos), and bisphenol A (found in containers, toys, and tin-can liners made from plastic). What we don't yet have, in many cases, are solid answers to the following questions (Groopman 2010): How long does a person have to be exposed to how much of a given chemical before cancer is likely to be observed? At a given level of exposure, what is the likely increased cancer incidence rate in a population? To what degree do benefits from exposure exceed risks? Industries that expose people to carcinogens exploit the fact that we often lack answers to such questions. Like cigarette manufacturers in the second half of the 20th century, they routinely hire experts to deny the scientific basis of alleged cancer risks and lobbyists to influence politicians to go easy on regulation (Egilman and Howe 2007).

This reality raises a public policy issue of gigantic proportions. You have learned that cancer incidence rates are increasing, that at least 90 percent of cancer cases have environmental causes, that research is, nonetheless, heavily skewed toward cure rather than prevention, that social inequality conspires against the effectiveness of efforts to encourage many individuals to minimize their exposure to cancer risk, and that industry is often reluctant to respond to plausible evidence of risk from known or suspected carcinogens. What, then, is to be done?

FROM PERSONAL HEALTH TO PUBLIC POLICY

> Medicine is a social science, and politics is nothing else but medicine on a large scale.
>
> —*Rudolf Virchow*

This startling sentence was written by the German founder of the science of cell pathology more than 150 years ago. It contradicts everything we think we know about medicine. We recognize that medical practitioners rely heavily on natural science to peer inside and fix the bodies of people who are ill. We understand that doctors do not, as a rule, look for cures in their patients' social relationships. Yet everything I have written about cancer points to the validity of Virchow's claim. It would be ridiculous to suggest that doctors should neglect treating the ill, but Virchow asks us to recognize that truly effective treatment must also involve investigating the social causes of disease and designing health and environmental policies that minimize disease risk. Your health is a deeply personal matter but, Virchow reminds us, it is simultaneously a pressing public issue.

Researching and writing this chapter led me to conclude that two types of public policies would go a long way toward reducing the cancer incidence rate in Canada. One type of policy is societal; the other is class-specific. Let me outline each in turn.

Societal Policies

Societal policies are not targeted at, but may benefit, specific categories of the population. Societal policies that would mitigate cancer risk in Canada include the following:

- *Requiring chemical safety tests and replacing highly toxic compounds.* Since 2007, the 27 countries of the European Union (EU) have required industry to prove that new chemical compounds are safe before exposing the public to them. In addition, manufacturers and importers of significant quantities of any chemical must file a report to the European Chemicals Agency demonstrating that they have adequately identified and managed risks linked to the chemical (European Commission 2010).

  Canadian industry bears no such responsibility. Before the government limits or bans the use of a chemical in Canada, it has to be proven dangerous. Of course, industry has little interest in voluntarily spending money to police itself, let alone forcing itself to use less toxic, but perhaps more expensive alternative compounds. Therefore, the Canadian process of testing and replacing toxic compounds is haphazard and slow. Our failure to regulate adequately the use of chemical compounds, including many that are reasonably anticipated to be carcinogenic, costs lives that could be saved if we followed the EU example.
- *Preparing a complete inventory of toxic waste sites and requiring their clean-up.* The government maintains a "Federal Contaminated Sites Inventory," that is, a list of known waste sites that pose a health danger to the public or the environment. However, the list of 7692 such sites (as of 2016) "does not include sites where contamination has been caused by, and which are under the control of, enterprise Crown corporations, private individuals, firms, or other non-federal levels of government (Treasury Board of Canada Secretariat 2016). It, therefore, seems likely that most of the country's toxic waste sites are not on the list and that we do not even know where all of our toxic waste sites are located. Nor do we systematically assign or take responsibility for their clean-up. This neglect takes an unknown toll on the lives of Canadians. Low-income Canadians undoubtedly suffer more than others do because known toxic waste sites are located in areas with a preponderance of low-income earners (Barlow and May 2000: 181–201).

- *Testing homes for radon and repairing inadequately sealed basements.* Radon is an invisible, tasteless, and odourless gas released by the natural decay of uranium in rocks and soil. It leaks into many homes through cracks in foundations and basement floors, loose pipe fittings, poorly insulated basement windows, and the like—problems that are especially common in the homes of the less well-to-do. According to the Canadian Cancer Society, about 2000 Canadians die from lung cancer caused by radon gas exposure every year (about twice as many as die from homicide and HIV/AIDS combined).

  In 2009, Health Canada cut by 75 percent the amount of radon considered safe in buildings. The World Health Organization (WHO) urges an additional 50 percent reduction, but even by Health Canada's generous standards, scientists estimate that in some parts of the country, including three provincial capitals (Winnipeg, Fredericton, and Halifax, with about 1.2 million residents in total), more than 20 percent of homes exceed the accepted maximum (Chen et al. 2008: 93). Unfortunately, we don't know which homes are at risk. We therefore need a comprehensive program to test for radon in every Canadian home and to repair houses that allow dangerous quantities of the gas to seep in.

A Class-Specific Program

Educational and screening programs teach people how to avoid cancer risks and detect cancer at an early stage, when it is most treatable. Such programs help, but they are significantly less effective among low-income individuals who may lack the education, access, money, and relatively unstressed work and family life that allow more highly educated and wealthier people to follow sensible and well-intentioned guidelines. Some anti-cancer policies must therefore focus on low-income people.

Sweden has established free neighbourhood "well-baby clinics" that bring medical care and social support into the community in such a way that inequalities in health status among children born into different social classes have been greatly reduced (Leon, Vågerö, and Olausson 1992). Similarly, the establishment of a string of wellness centres in low-income neighbourhoods across Canada could do much to equalize opportunities for good health and longevity in Canadian society.

Imagine a neighbourhood institution that offered tasty and healthful meals for the same price as a Big Mac, medium fries, and a Coke, with menus formulated to suit the ethnic tastes of area residents. Imagine also that this institution offered free nicotine patches and support groups for people wanting to quit smoking; free on-site cancer screening facilities; free counselling services for people facing marital or work-related stress; free competitive and instructional basketball, soccer, cricket, dance, and swimming activities for people of all age groups; and ads everywhere (multilingual, where necessary) featuring famous role models promoting the new centres. Although such facilities would, of course, cost billions every year, they would eventually lower healthcare costs by billions, all the while enhancing the quality and quantity of life.

A Political Choice

Politics influences who gets what, when, and how. It involves a struggle to divide resources among competing constituents, many of whom have compelling claims, but not all of whom can be completely satisfied, because resources are limited. Through elections, the expression of public opinion, contributions to political parties and other means, citizens and organizations have a say in who gets what, when, and how.

Some citizens and organizations—typically, those with more money—enjoy more political influence than other citizens and organizations do. As a result, public policy in Canada and elsewhere is inevitably biased in favour of the most influential. Conversely, as we have seen, Canadian health and environmental policies skew harm toward families and individuals with low incomes. The tilt of public policy thus has enormous consequences for who lives and who dies. This insight is what Virchow meant when he wrote, "politics is nothing else but medicine on a large scale."

Fortunately, money is not the only thing that enables people to gain political influence. Citizens routinely help shape public policy by organizing, mobilizing, petitioning, demonstrating, and voting.

A remarkable example is the movement to fight breast cancer. Neglected in research budgets until the 1980s, breast cancer research now accounts for a larger share of the Canadian cancer research budget than does any other form of cancer—more than three times the budget for research on prostate cancer, which has a 20 percent higher incidence rate (Canadian Cancer Research Alliance 2009: 30). The shift in funding priorities over the past three decades was due largely to political efforts by the women's movement to make breast cancer research a funding priority.

This case illustrates the rule: All decisions regarding public policy rest in the hands of the citizenry, including you and me. The anti-cancer policies I have recommended would cost a lot to implement. Putting them into practice would require making tough decisions about worthy alternatives. The proposed policies would inevitably hurt some well-established interests by setting new funding priorities. The good news is that they are achievable because they are political choices and worth the struggle because they would save so many lives.

CRITICAL THINKING QUESTIONS

1. The percentage of people who smoke varies by age, sex, and time period. Summarize these variations by consulting the Physicians for a Smoke-Free Canada (2009) site at http://www.smoke-free.ca/factsheets/pdf/prevalence.pdf. How do you explain these variations?
2. In your opinion, how effective would the policies outlined in this chapter be for reducing the cancer incidence rate? Why? Which categories of the population would likely support, and which would oppose, these policies? Why?
3. Do you think the second leading cause of death in Canada—heart disease—is subject to the same kinds of social forces analyzed in this chapter? Why or why not?

Aaron Millard

6
Gender Risk

WHAT IS GENDER RISK?

On the evening of December 16, 2012, Jyoti Singh Pandey, a 23-year-old physiotherapy student, went out with her boyfriend to see *The Life of Pi* in a suburb of Delhi, India. After the movie, the couple hopped on a minibus for the trip home. Six men were aboard, including the driver. They taunted the woman for going out in the evening. To them, she had failed to behave with appropriate modesty. The men beat the boyfriend unconscious with an iron rod and gagged and bound him. They then dragged Jyoti to the back of the bus, raping her repeatedly and penetrating her with the iron rod. They threw Jyoti and her boyfriend from the moving bus onto the road. The driver backed up, intent on killing her, but her boyfriend somehow managed to get her out of the way. Nonetheless, Jyoti died of her injuries on December 29.

About a week later, *The Times of India* reported Hindu spiritual leader Asaram Bapu as saying, "The victim is as guilty as her rapists" ("Delhi gang-rape victim" 2013; Kaul 2012).

The words of a religious authority like Asaram Bapu prevent us from dismissing Jyoti Singh Pandey's rape and murder as just the work of half a dozen sick individuals. When one man's belief is multiplied a million or a hundred million times over, it forms part of a culture, and for any element of a culture to endure, it must be supported by social arrangements that render it useful to some people. Bapu gave voice to a belief that is widespread, although far from universal, in India: Women deserve what they get when they fail to subordinate themselves to men and to behave with abject modesty.

To the degree that people like Asaram Bapu believe that Jyoti was in any way responsible for the tragedy that befell her, we must regard the crime against her as a phenomenon with deep social and cultural roots. Sexual violence, the neglect or murder of infant girls, disputes over dowries, domestic violence, and poor care of elderly women are responsible for the deaths of hundreds of thousands of girls and women in India every year (Anderson and Ray 2012; Harris 2013). Although mass protests against the treatment of women in India broke out in the days and weeks following Jyoti Singh Pandey's rape, many Indians see nothing wrong with the existing state of affairs.

What happened to Jyoti Singh Pandey is an expression of gender risk. **Gender risk** is the constellation of dangers associated with being a woman or a man (Hannah-Moffat and O'Malley 2007). Gender risk is not unique to India. In China, the Middle East, Africa, North America—everywhere, in fact—being a woman is associated with particular dangers. Of course, perils accompany manhood, too. For example, men are more likely than women are to earn their living in dangerous work settings, a circumstance associated with higher death rates. In this chapter, however, I focus on the gender risks of womanhood.

A remarkable feature of gender risk for women is its variability. Women face different gender risks to varying degrees, depending on the level of economic development, the cultural traditions, and the political circumstances of their country and region of residence. Describing and explaining some of that variability is the main goal of this chapter.

HIGH SEX RATIOS AS SIGNS OF TROUBLE

You will recall from Chapter 2 that the sex ratio is the number of men per 100 women in a population. At birth, the sex ratio for the entire world is 107, meaning that 107 boys are born for every 100 girls (see Table 6.1). Variation exists from one country to the next, with China and India having among the highest sex ratios at birth (112).

The sex ratio for people over the age of 65 looks much different. The elderly sex ratio for the world is just 80. While at birth there are 107 boys for every 100 girls, at the age of 65 and over, there are just 80 men for every 100 women. This fact demonstrates that, in all countries and on average, women live longer than men do. It is therefore astonishing to discover the magnitude of the world's highest

Table 6.1 Sex Ratios at Birth and at Age 65+ for Selected Countries

Country	Sex Ratio at Birth	Sex Ratio for People 65+
World	107	80
Canada	106	78
China	112	93
India	112	90
Qatar	102	171
United Arab Emirates	105	169

Source: Central Intelligence Agency (2016).

elderly sex ratios: 169 in the United Arab Emirates and 171 in Qatar. This means that, among elderly Emirati, for example, there are 169 men for every 100 women.

In general, a relatively high sex ratio indicates high gender risk for women. Thus, a high sex ratio at birth, which we find in India and China, suggests that the practice of **female-selective abortion** is widespread. Female-selective abortion involves terminating a pregnancy because the child is expected to be female. The practice is common in China and India, where the preference for sons is especially strong.

In China, the sex ratio began to rise above normal in the early 1980s. Two factors account for this: The technology for determining the sex of a fetus had become widely available, and the state had implemented a strict one-child policy for families as a means of controlling population growth (Li et al. 2007: 38).

In India, technology has affected the sex ratio as well. Prospective parents can now phone a clinic and, within hours, a technician will arrive at their doorstep with a portable ultrasound machine strapped to a motorbike to determine the sex of the fetus. There is then an assessment of whether further action is "required" (Sinha 2012). As a follow-up to ultrasound examinations, about 2000 female fetuses are lost in India every day.

Sex-selective abortion continues when Indian women emigrate. Thus, medical researchers studied 31 978 single-child live births to women born in India and living in Ontario between 2002 and 2007.

They found that the sex ratio for first children was 111, increasing to 136 for second children (Henry, Ray, and Urquia 2012).

According to the latest (2011) census, the sex ratio is continuing to rise in India (Arokiasamy and Goli 2012). With more men and fewer women in the population, men must compete more intensively for sexual partners and spouses, a situation that may be leading to an increase in male violence (Henrich, Boyd, and Richerson 2012). The kidnapping of girls and young women is also increasing. For example, in the Indian state of Andra Pradesh, an average of 22 girls and young women were kidnapped every day in 2012. Girls accounted for 70 percent of all child kidnappings (Khan 2012).

By comparing Indian sex ratios with those of highly developed countries such as Canada, economists have calculated how many Indian women are "missing" at each stage of the life cycle (Anderson and Ray 2012). About 12 percent of India's missing women are missing because of sex-selective abortion, with the remaining 88 percent dying in childhood, during their reproductive years, and at older ages.

Among the elderly, a relatively high sex ratio suggests that, throughout life, women die at a higher rate than men do because they experience more violence and inferior nutrition and health care compared to men in their own country and women in other countries. By this measure, some countries in the Persian Gulf region are the most dangerous for women.

It seems that most commentators would disagree, saying that the high sex ratio in some Persian Gulf countries is due to the large number of male migrant workers in the region ("UAE" 2002). Consider Abu Dhabi, the largest of the seven emirates in the United Arab Emirates. Non-citizens outnumber citizens four-to-one in Abu Dhabi. The sex ratio for citizens of all ages is 102 and for non-citizens of all ages is 250; for non-citizens of all ages in the labour force, the sex ratio is 556 (Statistics Centre—Abu Dhabi 2016). Clearly, the overwhelming majority of migrant workers are men. However, the work permits of nearly all migrant workers expire when the workers turn 60, at which time they must return home, mainly to India, Pakistan, Sri Lanka, the Philippines, and Bangladesh: "Expatriates working in the private sector who are over 60 years of age require the approval of the minister of labour. An approval may be granted if the worker is an expert or a consultant with expertise in a rare specialty" but manual and ordinary service workers must leave (Al Jandaly 2009). That is why there are almost no non-citizens in Abu Dhabi over the age of 64

(Statistics Centre—Abu Dhabi 2016). It follows that the high sex ratio at age 65 in the United Arab Emirates is not due to the disproportionately large number of male migrant workers in the country.

The map on the inside of the front cover provides a broader, global picture of women's safety, using a different index than the sex ratio. Based on the best recent data concerning rates of female homicide, rape, and domestic violence, it divides the world's countries into four categories, from those where women are physically safest (in green) to those where lack of physical security for women is most glaring (in red). The safest countries for women are in southwestern Europe and parts of Scandinavia. The most dangerous are in northern and eastern Africa, the Middle East, much of South Asia, and Peru (Kandiyoti 1988).

ECONOMIC DEVELOPMENT

How can we account for variation across countries in women's safety? After glancing at the map inside the front cover, one hypothesis springs immediately to mind. In general, rich, economically developed countries seem to be safer for women than poor, less developed countries are.

You can understand why this pattern often emerges by appreciating how economic development typically benefits women, at least in the long run. As the agricultural sector contracts and the manufacturing and service sectors expand, demand for women's paid work increases. Women are thereby drawn out of the home and into the system of formal education and the paid labour force. Glaring gender inequalities become evident. Men are paid more than women are for doing similar work. Domestic work, child care, and care of the elderly remain disproportionately the responsibility of women even after they start working full-time in the paid labour force. Women enter new social settings in which they may be exposed to violence. We see this phenomenon today in many industrializing countries, such as Mexico, Bangladesh, and Indonesia. There women move from the relative safety of their villages to factory towns, where they confront high rates of sexual harassment, violence, and homicide (Sarria 2009).

These and other issues galvanize many women to demand the vote, run for office, and champion state policies that promise to correct the problems they face. In short, economic development eventually increases women's independence and power, leading many women to achieve movement toward greater gender equality and safety.

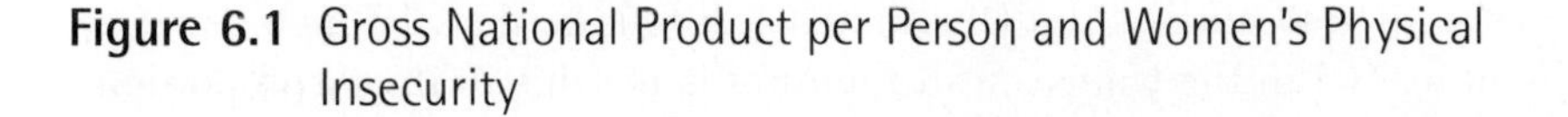

Figure 6.1 Gross National Product per Person and Women's Physical Insecurity

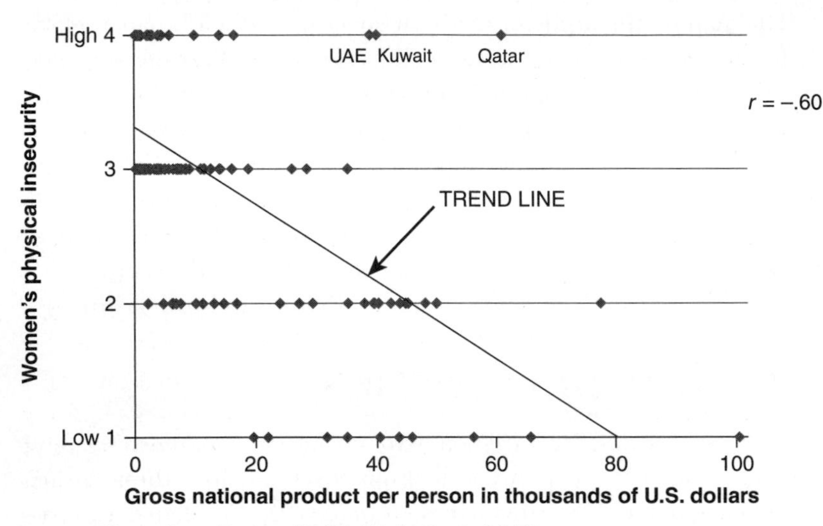

Sources: WomanStats Project (2013); World Bank (2013).

Figure 6.1 adds credibility to our hypothesis. It cross-classifies 137 countries for which data are available on two variables: (1) women's physical insecurity and (2) gross national product per person in thousands of U.S. dollars. "Women's physical insecurity" is a four-point scale in which a score of 1 is assigned to the safest countries (those with the lowest combined rate of female homicide, rape, and domestic violence), and 4 is assigned to the most dangerous countries (those with the highest combined rate). "Gross national product per person in thousands of U.S. dollars" is the U.S. dollar value of goods and services produced in a country in a year divided by the number of people residing in the country. Each country's score on both of these variables determines where its diamond-shaped marker is placed on the graph.

To interpret Figure 6.1, you need to recognize four things. First, some country markers fall above the graph's trend line while other country markers fall below it. If you were to calculate the distance between each country marker and the trend line, and then add up all those distances, you would find that the sum is smaller than the sum for any other straight line you could draw. The trend line is the best possible straight-line summary of the data in the sense that it minimizes the sum of the squared distances between each data point and the line.

The second thing you need to recognize about Figure 6.1 is that the trend line slopes downward. A downward-sloping trend line means that as one variable increases, the other decreases. (An upward-sloping trend line means that as one variable increases, so does the other.) In the present case, as gross national product per person increases, women's physical insecurity declines. Women tend to be safer in rich countries than in poor ones.

Third, notice the annotation "$r = -.60$" near the upper right corner of Figure 6.1. The value of *r* shows how strongly correlated the two variables are. The strongest possible correlations are -1.0 (for a downward-sloping trend line) and 1.0 (for an upward-sloping trend line). If the correlation in Figure 6.1 were -1.0 or 1.0, all of the country markers would fall right on the trend line. The weakest possible correlation is 0. If the correlation were 0, the country markers would be randomly scattered, many of them far from the trend line. Our correlation of $-.60$ indicates the existence of a strong, negative association between gross national product per person and women's physical insecurity.

The fourth and final feature of Figure 6.1 you need to recognize is that there are exceptions to the pattern we found. The biggest exceptions are the countries whose markers lie farthest from the trend line. These are countries in which gross national product per person has little bearing on women's safety. The three biggest exceptions are the United Arab Emirates, Qatar, and Kuwait, all in the Persian Gulf region. These countries are relatively rich, but a substantial number of women residing in them live in unsafe conditions. All three countries score 4 on the "women's physical insecurity" scale. Not surprisingly, therefore, survey data show that Kuwait is exceptional in terms of the degree to which its citizens are prepared to excuse rape or blame it on its victims, outpacing even India in this regard (Nayak et al. 2003).

We need a second hypothesis to account for exceptions such as Kuwait. One candidate is suggested by the fact that the three countries that Figure 6.1 identifies as most obviously breaking the overall pattern have Muslim majorities of more than 75 percent. This fact suggests that, independent of a country's wealth, women's physical insecurity may be related to the proportion of a country's population that is Muslim.

PATRIARCHY

Patriarchy is the system of inequality between women and men that exists in most societies. Strict patriarchy is characterized by male exclusivity in inheriting property, newly married couples establishing residence with or near the husband's parents, offspring assuming the surname of their father and tracing their lineage through the father's family, and the predominance of male authority inside and outside the family (Li et al. 2007: 38). Patriarchy is supported by structures of power rooted in the economy and the political system, and cultural traditions rooted in religion. For example, traditional Judaism, Christianity, and Islam are all patriarchal. They posit a male God, establish men as religious authorities, and recognize men as household heads.

Islamic doctrine is no more patriarchal than are the doctrines of other major religions. Moreover, enormous variation exists in the degree to which Muslims practise patriarchy. Some Muslims are feminists, insisting on equality between women and men in all spheres of life (Moghissi 1999; Yamani 1996). Some countries that are almost 100 percent Muslim, such as the Maldives and Tunisia, are relatively liberal when it comes to women's rights. (I discuss these cases later.)

On the other hand, some Muslim groups, notably the Taliban in Afghanistan and northwestern Pakistan, forbid women from receiving a formal education, working outside the home, or holding public office. They insist that, in public, women enshroud themselves in a *burqa*. They even plant bombs at girls' schools, throw battery acid in schoolgirls' faces, and gun down girls who advocate education because they are seen as threats to extreme patriarchy. Between 2006 and 2008, for example, 1153 attacks on educational targets (schools, teachers, and students) were recorded in Afghanistan, resulting in the death of 305 people, mainly girls and women (UNESCO 2010). From everything we know about the history of early Islam, it seems certain that the Prophet Muhammad would have been categorically opposed to the Taliban's restrictions on women and appalled by their acts of violence ("Women and Islam" 2013).

It is not widely appreciated in the West that many Muslim-majority countries have become more patriarchal only in the past 60 or 70 years.

To take just one example, in the 1950s, Alexandria, Egypt's second-biggest city, was so chic and cosmopolitan that it was known as Little Paris. Half a dozen languages were spoken on its streets. If you went to Sidi Bishr Beach in the summer, you would find it filled with tourists and affluent Egyptians of Arab, Jewish, and European descent listening to jazz and sipping cocktails. Women wearing bikinis socialized openly with men. Fast-forward to today, and Alexandria is a bastion of the conservative Muslim brotherhood and even more austere and patriarchal Salafists. Most women wear *niqabs* (full face covers), and at the beach they cover themselves head to toe. They wouldn't dream of wearing a skirt, let alone a bikini (Champion and Lagnado 2011; "Once Upon a Time" 2011).[1]

Such change in the way people dress signifies the spread and deepening of patriarchal practices in much of the Muslim world over the past half-century or so. Along with this entrenchment have come increased threats to women's safety, as we can see by briefly considering so-called honour killing.

Honour Killing

In 1992, the Shafia family left their native Afghanistan. After living in Pakistan, Australia, and the United Arab Emirates, the family finally settled in Montreal in 2007. Muhammad Shafia, the family patriarch, was a wealthy businessman. When his first wife, Rona Amir Muhammad, was unable to have children, he took a second wife, Tooba Yahya. The family could have been deported if the authorities had found out about the polygamous marriage so, in public, they referred to Rona as the aunt of the four children Tooba bore: Hamed, the eldest, a son, and daughters Zainab, Sahar, and Geeti.

Two years after arriving in Canada, the bodies of Rona (52), Zainab (19), Sahar (17), and Geeti (13) were found in a car submerged in the Kingston Mills lock of the Rideau Canal, 7 km (4 miles) north of Kingston, Ontario. In 2012, a jury found Muhammad, Hamed, and Tooba guilty on four counts of first-degree murder. According to the prosecution, the premeditated homicides were staged to look like an accident; the victims were first drowned and then placed in a car that was pushed into the canal ("Canadian Jury" 2012).

[1] I am grateful to my colleague, Rania Salem, for pointing out that the change is also due to the fact that the beach is no longer restricted to the wealthy.

Sidi Bishr Beach, 1959
Source: Elie Moreno

Sidi Bishr Beach, 2011
Source: travelpixs/Alamy Stock Photo

It emerged in trial that Muhammad was a domineering figure in the Shafia household and that Hamed was charged with keeping his sisters in line while his father was away on frequent business trips to Dubai. In the period immediately before the murders, Muhammad told Zainab she had to leave school for a year because she had a Pakistani-Canadian boyfriend. Terrified of her father, she fled to a shelter. The parents discovered evidence that Sahar was involved with a Christian boy: photos of her dressed in a short skirt, hugging him. They were horrified to find condoms in her room. Geeti was also in trouble. In open rebellion against the strict family regime, she began stealing, skipping school, failing classes, and wearing revealing clothes. She told a teacher she wanted to be put in foster care. Rona, the first wife, wrote in her diary that Muhammad beat her often, while Tooba referred to Rona as "the family servant."

The Shafia murders were considered honour killings. Among many people in the Middle East, North Africa, and parts of South Asia—the "patriarchal belt," as one sociologist calls it (Kandiyoti 1988)—honour is thought to depend on the sexual purity of female family members. Honour killings target women whose actual or suspected actions tarnish their family's reputation. The practice, while relatively infrequent, occurs in several religious communities, but is most common among Muslims (Abu-Odeh 1996; Ali 2003). A survey of 856 ninth graders in Amman, Jordan, found that about 40 percent of boys and 20 percent of girls believe that killing a daughter, sister, or wife can be justified if she has dishonoured the family (Eisner and Ghuneim 2013).

Of course, the overwhelming majority of patriarchal families seek to control female members merely through socialization and education, resorting, if necessary, to rules that restrict opportunities to deviate from family norms. Others employ physical punishment for transgressions. However, the United Nations estimates that, worldwide, about 5000 families a year go so far as to kill women who are thought to have dishonoured their family. Some experts in the field think that the real figure is three to four times higher because many honour killings are reported as accidental deaths (Wiseman 2010). Most honour killings occur in the patriarchal belt but, as the Shafia case illustrates, migration has made it a worldwide phenomenon (Gill, Begikhani, and Hague 2012; Korteweg and Yurdakul 2010; Patel and Gadit 2008). Since being raped is thought to dishonour the victim's family, on occasion a female rape victim is forced by her family to marry the man who raped her, thus restoring family honour and avoiding death.

Gender Bias in the Allocation of Food and Health Care

Honour killing is not the main source of excess female mortality in many Muslim-majority countries—sex bias in the allocation of food and health care is. One survey conducted in rural Bangladesh found that boys consistently consume more calories and protein than do girls, even after taking into account nutrient requirements due to varying body weight, activity levels, and other factors. Boys were also more likely to receive treatment at a free health facility than were girls (Chen, Huq, and D'Souza 1981). Sex-biased allocation of food and health care also seems to be chiefly responsible for the higher mortality rate of girls in Pakistan (Miller 2010). However, little such bias is evident in Morocco, Tunisia, and Jordan (Obermeyer and Cardenas 1997; Obermeyer, Deykin, and Potter 1993).

Figure 6.2 usefully summarizes this part of our discussion. It illustrates the association between the percentage of Muslims in a country and the four-point "women's physical insecurity" scale that, you will recall, combines rates of female homicide, rape, and domestic violence.

Figure 6.2 Muslims as a Percentage of Population and Women's Physical Insecurity

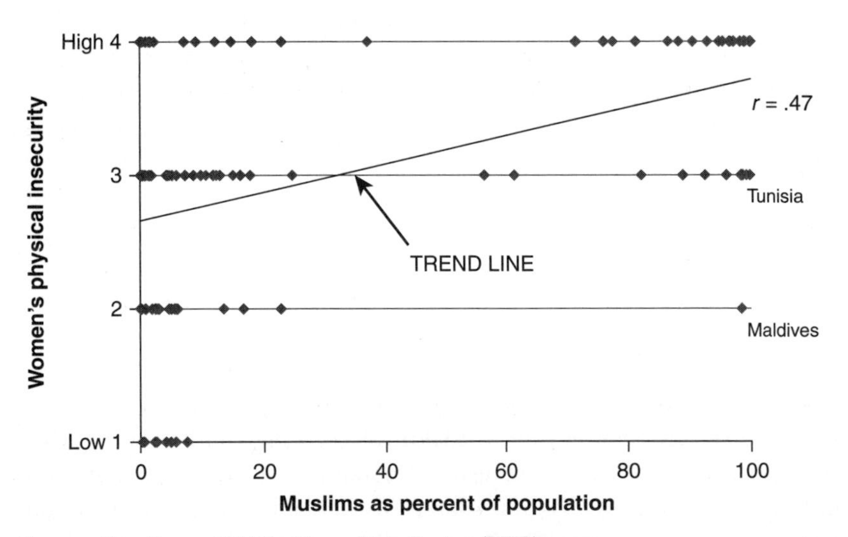

Sources: Pew Forum (2010); WomanStats Project (2013).

In brief, a strong tendency exists for women to be less safe in countries with proportionately larger Muslim populations ($r = .47$). Note, though, the existence of Muslim-majority countries, such as Tunisia and the Maldives, which do not fit this generalization particularly well. Such exceptional cases point to a third determinant of women's safety, apart from economic development and the percentage of a country's population that is Muslim: public policy.

PUBLIC POLICY

The Maldives is a tiny country consisting of a string of islands in the Indian Ocean (Library of Congress 2010; United Nations 2013a). More than 98 percent of its 350 000 inhabitants are Muslim, but it alone among the 32 Muslim-majority countries in Figure 6.2 scores 2 on the women's physical insecurity scale.

The country is no beacon of women's emancipation. Only 5 of its 77 members of Parliament are women, and it lacks a law against domestic violence. On the other hand, most women do not wear the veil, and they are not strictly segregated from men, although separate seating sections exist for women in public places such as stadiums and mosques. Women retain their maiden names after marriage. Inheritance of property is through both men and women, which is rare, if not unique in Muslim-majority countries. The proportion of women and men is approximately equal at all levels of the education system, and almost 56 percent of women over the age of 15 work in the paid labour force, the highest rate for women in the Muslim world and just 6 percent less than in Canada (United Nations 2013b: 356–57). The relatively high status of women in the Maldives is maintained as a matter of government policy.

Also exceptional is Tunisia. More than 99 percent of Tunisia's 10.7 million inhabitants are Muslim, yet Tunisia is among the 10 Muslim-majority countries in Figure 6.2 that score less than 4 on the women's physical insecurity scale. Again, government policy has a lot to do with that. In fact, even before mass protests overthrew the autocratic regime of Zine El Abidine Ben Ali in 2011, the government was ahead of much of the population in pushing for women's rights (Ben Salem 2010). Women have had the right to vote and hold office in Tunisia since 1959—by way of comparison, all Canadian women did not have the right to vote federally until 1960, when the franchise was finally extended to First Nations women.

An entire government ministry was created in Tunisia to ensure the equality of women and men and to encourage women to play a major role in the country's economic development. Consequently, young women are also educated as young men are. Women can move freely, open a bank account, and establish a business and a residence without the permission of their father or husband. Some 27 percent of judges and 31 percent of lawyers are women. Equality of women and men is affirmed in many Tunisian laws, including those governing the right to work and to give testimony equal to that of a man in court. (In many Muslim-majority countries, a woman's testimony in court is given only half the weight of a man's testimony.)

Still, by law, men continue to be favoured in inheritance, and just 25.5 percent of Tunisian women work in the paid labour force, about average for Muslim-majority countries (United Nations 2013b: 157). Most families remain strongholds of gender inequality. A 2007 survey found that about half of women and 37 percent of men believe that spouses should be equal and have the same authority in the family, but 32 percent of men and 20 percent of women absolutely opposed gender equality and strongly favoured patriarchy (Ben Salem 2010).

In 2011, the Islamist Ennahda party formed the first democratically elected government in Tunisia and in 2014 a peaceful transfer of power to a unity government took place. The country's new constitution enshrines some women's rights. Although Tunisia faces serious economic challenges and a worrisome terrorist threat, many analysts are cautiously optimistic that it will advance the gender equality agenda in the coming decades (Brym and Andersen 2016; Charrad and Zarrugh 2014).

The impact of government policy regarding women is registered by the level of gender inequality in a country: Where governments intervene most energetically to lower gender inequality, the level of gender inequality tends to fall. United Nations statisticians have usefully created an index of gender inequality for all 137 countries, as shown in Figure 6.3 (United Nations 2010). The index reflects gender-based disadvantages in three areas: (1) the labour market (the ratio of men to women over the age of 15 in the paid labour force); (2) empowerment (the ratio of men to women in Parliament and the ratio of men to women with secondary and postsecondary education); and (3) reproductive health (the percentage of women who die in childbirth and the percentage of adolescent women who have children).

The index ranges from 0, where women and men fare equally, to 1, where women fare as poorly as possible on all indicators. On this

Figure 6.3 Gender Inequality and Women's Physical Insecurity

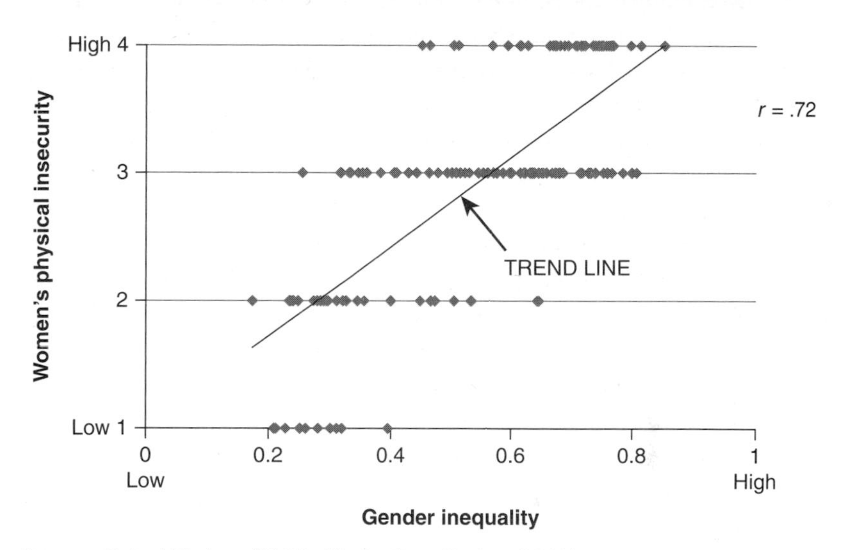

Sources: United Nations (2010); WomanStats Project (2013).

measure, the country with the lowest level of gender inequality is the Netherlands, with an index of .174. Canada ranks 17th lowest, with an index of .289. The country with the highest level of gender inequality is Yemen, with an index of .853. Figure 6.3 lets us see that gender inequality is strongly associated with women's physical insecurity (r = .72). Said differently, to the degree governments fail to pass laws promoting equality between women and men, women experience high levels of homicide, rape, and domestic abuse.

SUMMING UP

We have learned that women's physical insecurity is strongly associated with a country's level of economic development, the percentage of its population that is Muslim, and the degree to which its government passes laws aimed at promoting gender equality. The trouble is that the three variables associated with women's physical insecurity are correlated with one another. For example, Muslim-majority countries tend to be poor and lack governments that promote gender equality. Rich countries tend to have small Muslim minorities and governments that promote gender equality.

These facts imply that we can't easily determine whether each variable is significant by itself. For example, if we find that women's physical insecurity is greater in Muslim-majority than in Muslim-minority countries, this may have nothing to do with Islam; it may be solely the result of the fact that Muslim-majority countries tend to be less economically developed than Muslim-minority countries are. Moreover, the fact that all three of our variables are correlated means that we cannot easily determine the relative importance of each variable and the combined effect of all three variables on women's physical insecurity.

A widely used statistical procedure allows us to figure out the unique and combined effects of the three variables on women's physical insecurity. It is beyond the scope of a chapter intended for first-year college and university students to offer such an analysis in full, but the following points can be made. First, it turns out that all three factors contributing to women's physical insecurity have independent or unique effects that are almost certainly not due to chance. Second, government policy promoting gender equality has the strongest effect on women's physical insecurity. The percentage of a country's population that is Muslim has an effect that is about 50 percent weaker. Economic development has an effect that is about 25 percent weaker still. Third, all three factors combined account well for variation in women's physical insecurity across countries.[2]

INTERSECTIONALITY

Canadian women live in safety compared to women residing in countries where governments do little to reduce gender inequality, especially if the countries have a Muslim majority and are poor. Nonetheless, the safety of Canadian women is only relative.

[2] For readers familiar with multiple regression analysis, coefficients are shown below. Additional analysis revealed a normal distribution of standardized fitted values of the dependent variable plotted against the standardized values of the residuals. No evident pattern in the scatterplot suggested a misspecified model.

	b	*s.e.*	*beta*	*t*	*p*	Cumulative adjusted R^2
Constant = 1.615		.272		5.940	<.001	
Gender inequality index	2.362	.430	.487	5.497	<.001	.516
Percentage Muslim	.006	.001	.252	4.163	<.001	.561
GNP per capita	−9.117E−006	.000	−.191	−2.268	<.05	.574

In 2009, about 677 000 Canadians (24 of every 1000 Canadians over the age of 14) reported that they had been sexually assaulted in the preceding year. Seventy percent of the victims were women (Perreault and Brennan 2010). Between 2000 and 2009, 738 spousal homicides took place in this country. Women accounted for about three-quarters of the victims. In 2009, women reporting spousal violence to the police were about three-and-a-half times more likely than men were to report being sexually assaulted, beaten, choked, or threatened with a gun or a knife (Statistics Canada 2011).

While rates of spousal violence and spousal homicide have been stable in recent years, reported dating violence is on the rise. It accounts for more than one-quarter of all violent incidents reported to the police and almost one-third of all homicides perpetrated by intimate partners. More than 80 percent of the victims of police-reported dating violence are women (Mahony 2010).

Although I have used such nationwide figures on violence against women throughout this chapter, it is important to note that, in doing so, I have neglected variation *within* countries, which is always substantial. In particular, violence against women in lower classes and in disadvantaged ethnic and racial groups is higher than is violence against women in upper classes and in privileged ethnic and racial groups, both in Canada and in other countries (see, for example, Romero 2013). Where lower class position and disadvantaged ethnic/racial status intersect, violence against women is highest. This phenomenon is known as **intersectionality**, the tendency for class and ethnicity/race to amplify the effect of gender on a range of behaviours, including violence (Choo and Ferree 2010).

The impact of intersectionality in Canada is evident if we consider the plight of Aboriginal women. On average, Aboriginal women occupy a relatively low class position and are members of a highly disadvantaged racial group. They are about three times more likely than non-Aboriginal women are to live in a dwelling that requires major repairs, more than twice as likely to be unemployed, and nearly twice as likely not to complete high school. On average, they earn about 25 percent less than non-Aboriginal women do (O'Donnell and Wallace 2011).

Conquest, colonization, punishing government policy, and widespread racial discrimination have pushed Aboriginal women into situations associated with a heightened risk of violence—overcrowded dwellings, lack of access to women's shelters and other government services, high rates of alcohol abuse, disproportionate involvement in street prostitution, and so on (Amnesty International 2009; King 2012). Consequently, Aboriginal women suffer five times the homicide rate of non-Aboriginal women and experience two-and-a-half times the rate of spousal violence. Those who experience spousal violence are 41 percent more likely than non-Aboriginal women are to sustain an injury (O'Donnell and Wallace 2011).

It is often said that the living conditions of Aboriginal Canadians are not much different from the living conditions of many people in less-developed countries. The impact of intersectionality on violence against Canada's Aboriginal women adds weight to that assertion.

In sum, although Canadian women live in safety relative to women in most other countries, women's physical insecurity in Canada is still substantial. Moreover, as the situation of Aboriginal women demonstrates, for Canadian women in lower classes and disadvantaged ethnic or racial groups, physical insecurity is much higher than the national average suggests. If this is what relative safety for women looks like, it is tempting to despair when thinking about the dangers facing women in, say, India and Pakistan.

THE IMPORTANCE OF ACTIVISM

Yet despair is precisely the opposite of what is needed to deal with women's physical insecurity worldwide. One 36-nation study found that the more vigorous and independent the women's movement in a given country, the greater the prominence of women's safety in that country's political agenda (Weldon 2002). Without demonstrations, petitions, television appearances, newspaper articles, blogs, and academic research bringing violence against women to the attention of the public and into the realm of public debate, the problem remains a political non-issue.

The case of 16-year-old Pakistani schoolgirl, Malala Yousafzai, stands as a symbol for a generation. A resident of the Swat Valley, a Taliban stronghold, she began promoting the education of women and women's rights at the age of 11. She rose to such prominence that she won Pakistan's first National Youth Peace Prize and earned nominations for the international Children's Peace Prize and the Nobel Peace Prize. Because of her activities, a Taliban gunman tried to assassinate her in 2012. Shot in the head and the neck, she has since managed to recover almost completely and is continuing her heroic work (Walsh 2012).

Magnifying the impact of a strong women's movement is the establishment of government departments able and willing to respond positively to activism. By funding, coordinating, providing legal and other expert support, and opening up access to the policy-making process, such institutional mechanisms multiply the effectiveness of a strong women's movement.

Between 1967 and 2006, Canada was a world leader in developing programs to deal with women's safety issues because it enjoyed a relatively strong women's movement and comparatively responsive government institutions (Weldon 2002: 154–63). Then, from 2006 to 2015, the Conservative government removed long-standing targets for increasing gender equality in the country (Bell 2013). The election of a Liberal government in 2015 signified reversion to a gender equality agenda; women hold 15 of 31 cabinet positions in the current Trudeau government. Nonetheless, the Conservative interlude highlights the fact that no advance can be taken for granted and that much work remains to be done in Canada, as elsewhere, to make society as safe for women as it is for men.

CRITICAL THINKING EXERCISES

1. The degree to which the Canadian government promotes gender equality declined between 2006 and 2015. Do a Google search using the keywords "Status of Women Canada." Based on the documents you find, write a 500-word essay outlining the relevant policy changes implemented between 2006 and 2015. In a concluding paragraph, explain how these changes might influence women's safety in this country.
2. Evidence presented in this chapter strongly supports the view that nothing inherent in Islam prevents gender equality in Muslim-majority countries. A report by the United Nations, *The Arab Human Development Report 2005: Towards the Rise of Women in the Arab World*, details how gender equality might be achieved in the Arab world. Review the report, found at http://hdr.undp.org/sites/default/files/rbas_ahdr2005_en.pdf, and in a 500-word essay, outline its main recommendations for increasing gender equality. In a concluding paragraph, recommend additional policies for promoting women's safety.

Aaron Millard

7

Canadian Genocide[1]

"NOW YOU ARE NO LONGER AN INDIAN"

Richard Cardinal was born in Fort Chipewyan in northern Alberta. He was **Métis**.[2] One day in 1971, he and his siblings were abruptly removed from their parental home by child welfare authorities and placed in separate foster homes. None of the foster parents were of Indigenous origin and none lived in Fort Chipewyan, so all of the Cardinal children were suddenly uprooted from everything they knew. Richard was four years old.

Over the next 13 years, Richard lived in 16 foster homes and 12 shelters, group homes, and other facilities—more than two changes of residence per year on average. Often abused, he felt abandoned in an alien world. He was angry but rarely showed outward signs of his torment except for persistent bedwetting. Largely because of that problem, one set of foster parents after another gave up on him. One couple had a good idea. They would be willing to keep him if the social worker could supply them with a rubber sheet for Richard's bed; they were unable to find one in their rural community. The social worker said she couldn't help, so Richard was sent elsewhere.

Richard was a sensitive and intelligent boy and a talented writer. We know this because he wrote a diary that was excerpted in the *Toronto Star* and became the basis for a short film (Obomsawin 1986). Richard never got to see the film, however, because in June 1984 he hanged himself from a birch tree in the backyard of his last foster home. He was 17.

[1] I am grateful to Jeff Denis and Howard Ramos for critically commenting on a draft of this chapter and rescuing me from errors of fact and interpretation.
[2] Métis are of Indigenous and European (usually French) origin. They reside mainly in the western provinces and Ontario.

The details of Richard's case are unique but its general characteristics are common. By the mid-1980s, about 20 000 Indigenous children had become wards of the provinces. They were part of the so-called "Sixties Scoop" that saw Children's Aid Societies remove neglected and abused children from their families. ("Sixties Scoop" is a misnomer because it went on for about 25 years.) They were typically forced to live with white families in communities that were not within visiting distance of their siblings. Some victims of the "scoop" were taken from relatives who were happy to raise them and some were sold to adoptive families in the United States for as much as $30 000 each, with the proceeds going to provincial child welfare agencies (Carreiro 2016).

At the same time, some 150 000 Indigenous children had lived or were living in residential schools (Milloy 1996; Truth and Reconciliation Commission 2012). The schools were funded by the federal government and run by the Roman Catholic, Anglican, United, Presbyterian, and Methodist churches. Their aim was to help Indigenous children "escape" their families and their cultures, facilitating their assimilation into white Canada. As one of the staff members at Saskatchewan's File Hills School proclaimed after he sheared off young Charlie Bigknife's hair, "Now you are no longer an Indian" (quoted in Truth and Reconciliation Commission 2012: 22). Food was meagre, discipline severe. Staff members often subjected students to emotional, physical, and sexual abuse. Between 1942 and 1952, some children were knowingly deprived of certain basic nutrients as part of a series of experiments conducted by the country's leading nutritionists in cooperation with the federal government (Mosby 2013). In total, about 6000 students died while in the "care" of residential school staff members. For a first-hand account of life in a residential school, please see this chapter's appendix.

Some Canadians now say that the wholesale removal of Indigenous children from their families amounts to a form of "cultural genocide." My view is that this modifier makes light of the wrongdoing. It was not just a set of cultures that European colonists and their descendants were killing off.

GENOCIDE

Resolution 96(1) of the First General Assembly of the United Nations declares that **genocide** is "a denial of the right of existence of entire human groups." It goes on to say that genocide is a crime that occurs "when racial, religious, political and other groups [are] destroyed, entirely or in part" (United Nations 1946: 188–89).

Most historians agree that the first modern genocide was the Turkish slaughter of about 1.5 million Armenians between 1915 and 1923 (Armenian National Institute 2016). The Turks killed approximately 75 percent of the Armenians in the Ottoman Empire. The 20th-century largest genocide was the Nazi extermination of 6.1 million Jews during World War II. The Jewish population of the countries occupied or dominated by the Nazis was reduced by just over 73 percent between 1939 and 1945 (Lestschinsky 1948: 60). It seems reasonable to use these percentages as points of reference for judging whether European colonists and their descendants committed genocide against Canada's Indigenous people.

Sustained contact between Indigenous people and European settlers began about 500 years ago. Most demographers and anthropologists who have studied the subject believe that, at the time, between 350 000 and 500 000 Indigenous people lived in the area we now call Canada (Aylsworth and Trovato 2015 [2006]). Let us assume that 425 000, the midpoint between these two figures, is the best estimate. In 1911, the Canadian census reported a mere 105 611 people of Indigenous ancestry. By then, the Indigenous population had undergone a reduction of about 75 percent since the first sustained contact—proportionately about the same as the collapse of the Armenian population in Turkey and the Jewish population in Nazi Europe.

American poet E. E. Cummings once wrote that "nothing recedes like progress" (quoted in Welch 2016). We can appreciate the sense of his epigram by noting that, in the case of Canada's Indigenous people, the mechanisms of population decline were not as purposefully designed, systematically organized, and technologically "advanced" as those of the later Turkish and especially Nazi killing machines (Bauman 1991 [1989]). Although there was violence, there were no equivalents of the gas chambers or the *Einsatzgruppen*, paramilitary death squads that roamed the Eastern European countryside rounding up and killing Jews. Nor was there an outright policy of extermination in Canada. Consequently, population decline among Indigenous Canadians took four centuries to play out, as opposed to just eight years in the Armenian case and a mere six in the Jewish case.

Nonetheless, the conclusion seems to me inescapable that between the early 1500s and the early 1900s, Indigenous Canadians were denied the "right of existence" by the European colonizers of what was to become Canada, resulting in their destruction, "entirely or in part." Specifically, Canada's Indigenous population circa 1911 was about one-quarter the size it was four centuries earlier because of "disease, starvation and warfare *directly stemming from European settlement and practices*" (Aylsworth and Trovato 2015 [2006]; my emphasis).

A few Indigenous people, such as the Sadlermiut **Inuit**[3] on Southampton Island in the Arctic and the Beothuk in Newfoundland, suffered complete extinction. Year-round European settlement began in Newfoundland in the 1600s. The Europeans viewed the Beothuk as a nuisance, so they paid some Mi'kmaq from Nova Scotia to help kill them off. The surviving Beothuk, numbering fewer than 1000, gradually withdrew from European contact. Reduced to a tiny refugee population living off the meagre resources of the Newfoundland interior, they soon died of starvation and disease. Today, all that remains of the Beothuk aside from their tragic history and a few museum artifacts is a statue outside the Newfoundland and Labrador legislature in St. John's (Pastore 1997).

Although there was some violent conflict between colonists and Indigenous people almost from the beginning—in Atlantic Canada, the Mi'kmaq allied with French settlers (Acadians) to resist the British for nearly a century—initially there was also much cooperation. Indigenous people exchanged valuable furs for blankets, iron tools, rifles, and other Western products, guided the newcomers into the interior, and served as reliable military allies (Innis 1977 [1930]; Miller and Parrott 2015 [2006]). Recognizing their utility, the British even promised to support an independent confederacy of Indigenous people known as the Indian Buffer State west of the Great Lakes and east of the Ohio River.

However, the relationship changed for the worse in the early 1800s. International demand for furs declined. After many Indigenous people helped the British win the War of 1812 against American invaders, it became clear that they were no longer of much military use. Besides, the British now had western settlement in mind for the rapidly expanding population in the motherland. Already by 1763 they had passed laws allowing them to take possession of Indigenous land. Now they abandoned their commitment to an independent Indigenous confederacy and

[3] The Inuit are Indigenous Canadians who reside in the country's northern regions.

adopted a more or less consistent pattern of treatment across the territory that later become Canada:

- To make way for white agriculture, industry, transportation, and settlement, the state concentrated most Indigenous people on "reserves," often by force. In northwestern Ontario, for example, whites flooded Indigenous rice paddies, depleted sturgeon populations, imposed hunting restrictions, and legally extinguished one Indigenous community and forcibly relocated six others to reserves. The courts justified the chaos that whites brought on the Anishinaabe by labelling the latter "heathens and barbarians ... an inferior race ... in an inferior state of civilization" (quoted in Denis 2015: 224). As a result of similar actions throughout Canada, many reserves had meagre resources, including traditional food supplies. In the west, the bison, a mainstay of Indigenous food, clothing, and shelter, was largely killed off to facilitate colonization.
- The reserves were characterized by poor, crowded housing, inadequate sanitation, and unsafe drinking water—ideal conditions for the spread of disease. Compounding the problem was that, in their contacts with settlers, Indigenous people continued to be exposed to smallpox, measles, influenza, and tuberculosis, "European" illnesses to which they had little resistance, especially in their often-malnourished state.
- The government expended minimal effort to prevent unscrupulous white traders, mainly Americans, from exploiting the Indigenous population by exchanging whisky for furs. It was only in 1874 that a mere 309 Mounties were sent out to police the entire prairie area: one Mountie on horseback per thousands of square kilometres. Destitution and loss of tradition combined with excessive alcohol consumption to fuel child neglect and abuse, wife abuse, and suicide, all largely unknown before sustained contact with Europeans, as the late Memorial University anthropologist Jean Briggs showed in her classic study of the Inuit (Briggs 1970).

SEEING RED

Given these conditions and the resulting population loss, most whites writing on the subject in the 1800s believed that Canada's Indigenous people were doomed to extinction (Francis 2011 [1997]: 31–74). This fact suggests that many white settlers understood that their settlement practices had deadly effects. They could live with this knowledge, which should have been devastating, only by developing certain ideological defences.

In particular, many white settlers created an image of Indigenous people as inherently dangerous. They also came to believe that Indigenous people themselves were the cause of their own sorry state. These rationalizations helped white people ease their minds as they went about their deadly business. For if Indigenous people posed a threat, any action against them could be regarded as defensive and therefore justifiable. And if Indigenous people were responsible for their own pitiful condition, then whites could not reasonably be blamed for their demise. Let us consider each of these notions.

Before the War of 1812, many whites regarded Indigenous people as "noble savages"—strong, independent, living in harmony with nature and lacking the corrupting influences of Western civilization. However, when Indigenous people became less useful to the settlers in the years after the war, many whites started emphasizing their inherent bloodthirstiness and propensity for violent drunkenness. *The History of the Dominion of Canada*, a book widely used in schools throughout the country, described Indigenous people as follows: "cruel," "rude," "false," "crafty," "savages," "torturing," "of unclean habits," "without morals" and "ferocious villains" who plotted against the Europeans with "fiendish ingenuity" (Clement 1897: 12–13). White Canadians also came to see Indigenous people as impediments to the advance of Western civilization. In short, Indigenous people were increasingly regarded as a danger, either to life and limb or to progress itself.

These sentiments were propagandized in hundreds of short stories, novels, and movies about the North-West Mounted Police, later renamed the Royal Canadian Mounted Police. In these works of fiction, which were voraciously consumed by the Canadian public until the late 1930s, the consistent aim of the Mounties was to protect pioneers from the murderous enemies of white society. To borrow Vancouver historian Daniel Francis's metaphor: Just as the forests had to be cleared of trees to make way for farmers' fields and the prairie sod had to be broken so farmers could plant their crops, the lawless savages had to be pacified. Only after the frontier was made safe could the railways be built and the settlers feel at home (Francis 2011 [1997]: 95). It never occurred to the authors of these dreams that Indigenous people might have a case for opposing the march of Western civilization insofar as it was destroying their way of life and killing them by the thousands.

Victim blaming involves holding the targets of wrongful action responsible for the harm that befalls them. It became popular in the 1800s and is still widespread today. A nationally representative survey conducted in 2016 found that 32 percent of non-Indigenous Albertans, 35 percent of non-Indigenous Manitobans, and 41 percent of non-Indigenous people from Saskatchewan think the biggest obstacle to economic and social equality between Indigenous and non-Indigenous Canadians is the Indigenous people themselves (Environics Institute 2016: 22). A similar survey conducted in 2013 found that 60 percent of Canadian adults think that Indigenous Canadians' problems are self-inflicted (Mahoney 2013). Little trace of historical responsibility for the plight of Indigenous Canadians can be found among people who hold such views.

A recent study shows how victim blaming has persisted even in unexpected places. McMaster University sociologist Jeff Denis spent 18 months living in Fort Frances, a mill and mining town of about 8000 residents in northwestern Ontario. During his stay, he interviewed and systematically observed the interactions of white and Indigenous residents (Denis 2015).

An established school of thought in social psychology known as the **contact hypothesis** holds that prejudice will be reduced to the degree that majority and minority group members engage in friendly, informal, personal interaction (Pettigrew and Tropp 2006). Denis found a high level of such interaction and even intermarriage between white and Indigenous residents of the Fort Frances region.

The contact hypothesis would lead us to expect little victim-blaming and other forms of prejudice among non-Indigenous people with close Indigenous friends. However, Denis found the opposite. For example, Denis asked white people to think about the main reasons why Indigenous people are more likely than non-Indigenous people to be poor. Fifty percent of all of his respondents, and 50 percent of his respondents with close Indigenous friends, placed all of the blame on Indigenous people themselves. Twenty-four percent of all of his respondents, and 25 percent of his respondents with close Indigenous friends, placed half the blame on Indigenous people themselves. Thus, close Indigenous friendships had no effect at all on victim-blaming.

Not only did most whites, including those with close Indigenous friends, blame Indigenous people themselves for their social problems, they rejected policies designed to improve Indigenous living conditions and resented people who fought for Indigenous rights. Many of these whites were perfectly friendly and even on intimate terms with Indigenous people. However, they insisted that the subordinate position of Indigenous people in the social order was appropriate given their behaviour and that improvements in their social standing should come about only through the efforts of Indigenous people themselves, regardless of the many impediments thrust on them by the history of white settler colonialism in Canada.

INEQUALITY

Significantly, until the 1940s, the Canadian government offered only one clear path to survival for Indigenous Canadians, and it followed the same logic that Denis found among most of the whites he interviewed in Fort Frances. The government decided that Indigenous people could live and live well as individuals—if they were willing to give up their identity and their culture, their way of life. To that end, efforts were made to convert the heathen to Christianity and transform the hunter and the trapper into a farmer or a labourer. The 1876 Indian Act officially assigned the government the task of integrating Indigenous people into mainstream Canadian society. Shortly thereafter, the residential school system became part of that grand design.

We can gain a sense of how well the plan worked by first comparing the position of Indigenous people today to the position of all Canadians. Canadians often decry the race problem in the United States and think it is worse than the Canadian race problem. Therefore, we can further improve our appreciation of the depth of Canada's Indigenous problem by comparing the standing of Indigenous people in Canada to the standing of African Americans in the United States. Table 7.1 makes both comparisons.

The first column in Table 7.1 shows that the standing of Indigenous Canadians is far below that of all Canadians on all indicators. They are about twice as likely to be unemployed and ten times as likely to be in prison. They are around six times more likely to commit homicide. Their **infant mortality rate** (the number of babies who die before the age of one per 100 000 live births) is more than twice as high. They earn only 60 percent of the Canadian median income and graduate from university at just 37 percent the rate of all Canadians. In Canada, **life expectancy** (the average number of years a person can be expected to live) is about 82 years. For Indigenous people it is less than 75 years.

Table 7.1 Inequality Indicators: Indigenous Canadians and African Americans

Indicator	a Indigenous Canadian	b African Americans	c (a/b)
1. Unemployment rate in percent versus national rate	2.1	1.9	1.11
2. Rate of imprisonment per 100 000 population versus national rate	10	3	3.33
3. Homicide rate per 100 000 population versus national rate	6.1	3.7	1.65
4. Infant mortality rate per 100 000 live births versus national rate	2.3	2.0	1.15
5. Median income in dollars as a percentage of national median	60	74	0.81
6. Percent with university degree versus national percentage	37.0	65.5	0.56
7. Life expectancy in years versus national life expectancy, expressed as a percentage	91	95	0.96

Sources: BlackDemographics.com (2016); Gilmore (2015); Statistics Canada (2015a).

To see how Canada's biggest race problem compares with that of the biggest race problem in the United States, we now turn to the second column in Table 7.1, which measures the standing of African Americans relative to all Americans. Dividing the figures in the first column by the figures in the second column gives us the comparison we are after. The third column in Table 7.1 shows that inequality between Indigenous Canadians and all Canadians is more severe than inequality between

African Americans and all Americans. For instance, African Americans are three times more likely to be in prison than are all Americans, but Indigenous Canadians are ten times more likely to be in prison than are all Canadians. While African Americans graduate from university at more than 65 percent the rate of all Americans, Indigenous Canadians graduate from university at just 37 percent the rate of all Canadians. Relatively speaking, Indigenous Canadians are significantly worse off than African Americans are. To varying degrees, most of them are marginalized in, not integrated into, Canadian society.

Note that Table 7.1 is based on recent statistics. This means they represent the situation more than seven decades after relations between Indigenous and non-Indigenous Canadians entered a phase when Indigenous organizations, movements, and individuals started demanding justice, public opinion became more sympathetic to the plight of this country's first inhabitants, and the government started working in earnest to improve their prospects. Seen in this context, the figures in Table 7.1 reveal just how agonizingly slow progress on this front has been.

Improvement in relations came about for numerous reasons (Denis 2016: 7–9):

- During World War II, about 4000 Indigenous people volunteered for army service. They earned the respect and admiration of many non-Indigenous Canadians, and when the soldiers returned home, they sought social, economic, and political change that would improve the lives of all Indigenous Canadians.
- Indigenous leaders formed political organizations to pressure the federal government to remove coercive regulations and weaken the push to forced assimilation.
- International bodies began calling on Canada to do a better job of respecting human rights and curbing discrimination.
- The government began consulting Indigenous leaders on policy matters.
- In 1960, **Status Indians** won the right to vote in federal elections.[4]

[4] Status Indians are Indigenous people who live throughout the country and are registered as "Indians" under the Indian Act. Most of them belong to a band that signed a treaty with the Crown. **Non-Status Indians** are people who were once Status Indians but lost the title. For example, until 1985, female Status Indians automatically lost their status if they married a non-Status Indian. Status and non-Status Indians are commonly called **First Nations**. About a quarter of First Nations people are non-Status Indians (Statistics Canada 2015b).

Figure 7.1 Indigenous Population, Canada, 1901–2011

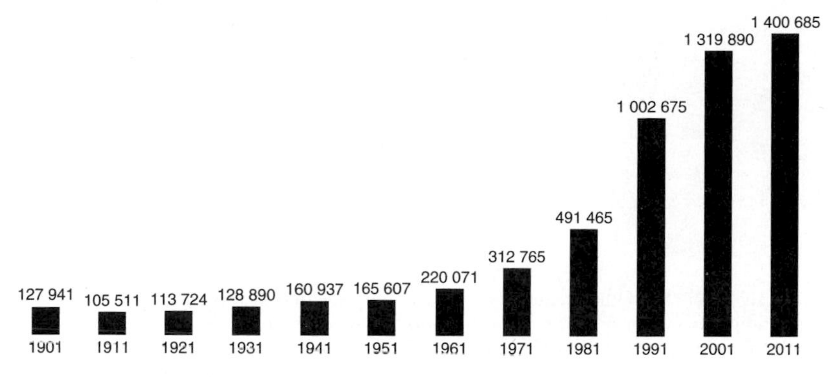

Note: The compulsory national census was replaced by a voluntary *National Household Survey* in 2011. Some census subdivisions were excluded from final estimates because responses were too few and estimates were therefore unreliable. For example, in Saskatchewan, where about 16 percent of the population is Indigenous, only 57.4 percent of the province's subdivisions had usable data (Brym 2014:9). As a result, the 2011 figure in this graph is probably an underestimate.

Source: Statistics Canada (2015b, 2016).

- The government began closing down residential schools (a process it did not complete until 1996).
- The federal budget, staff, and facilities for Indigenous health care increased phenomenally (Young 2015 [2006]). In the space of little more than a decade, and largely as a result of this intervention, a demographic transition took place. The infant mortality rate fell. Life expectancy increased. In the 1950s, for the first time since first contact, the size of the Indigenous population started to rise significantly. In addition, more people of Indigenous origin started to identify as such in the census. By 2011, about 1.4 million Canadians claimed to be of Indigenous ancestry, around the same number who claimed to be of Italian or Chinese origin (see Figure 7.1).

CULTURE, CLASS, AND GENDER

One might get the impression from our discussion thus far that Indigenous Canadians form a socially homogeneous group. They do not. They are differentiated along multiple axes, including culture, economic standing, and gender. Let us touch on each of these dimensions in turn.

Figure 7.2 Indigenous Population by Self-Identification, Canada

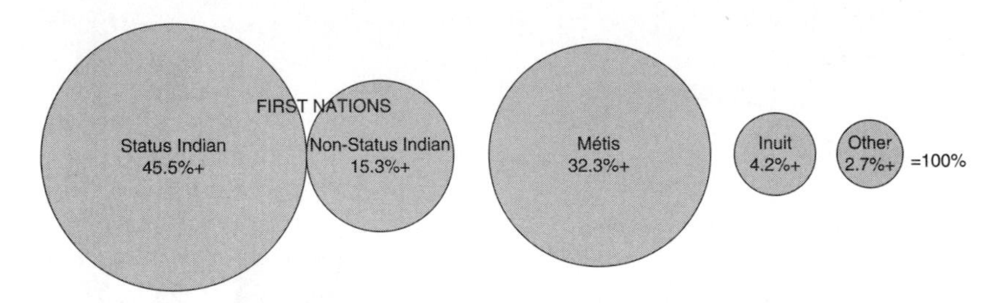

Note: Except for the "Other" category, all people represented in this chart reported a single Indigenous identity in the 2011 *National Household Survey*. The "Other" category includes people who reported multiple Indigenous identities or declared Indigenous identities not included elsewhere.

Source: Statistics Canada (2015b).

Nearly a third of Indigenous Canadians are of Métis ancestry and more than 4 percent are Inuit. About 61 percent are First Nations, that is, Status and non-Status Indians. Each of these broad groupings is itself culturally diverse. For instance, the First Nations are made up of more than 600 bands speaking more than 60 languages (Environics Institute 2010; Statistics Canada 2015b; Voyageur and Calliou 2000–01; Figure 7.2).

The Indigenous population is also stratified economically. On average, the earnings of Métis and urban First Nations people are higher than those of First Nations people who live on reserves (Denis 2016: 16). And within each of these groupings, we again find considerable variation.

At one extreme, about a third of First Nations children live in low-income families, defined as families that earn less than half the median income of all Canadian families (House of Commons Canada 2010: 31).

The Indigenous middle class has grown significantly in recent decades. For example, in 1950, only about 200 businesses were owned or operated by members of First Nations. Today, that figure is more than 30 000. British Columbia's first Indigenous lawyer was called to the bar in 1962. Today, there are about 200 Indigenous lawyers in the province (Griner 2013; Law Society of British Columbia 2015).

Then there are the wealthy. Montreal Canadiens goalie Carey Price, whose mother is former chief of the Ulkatcho First Nation in British Columbia, earns US$7 million a year ("Carey Price" 2016). Dave Tuccaro, a member of the Mikisew Cree First Nation in Fort Chipewyan, Alberta, got rich off the oil sands. Starting as a heavy equipment operator in the 1980s, he established companies that provide the oil industry with everything from heavy hauling to laboratory services. His net worth in 2012 was more than $100 million (Vanderklippe 2012). In general, communities that do business with the oil and gas industry have prospered. Foremost among them is Fort McKay, the richest Indigenous community in Canada, with an annual household income of $120 000 (Bakx 2016). In 2013, Jim Boucher, chief of the Fort MacKay First Nation (population: 826), earned $644 441. That same year, Ron Giesbrecht, chief of the British Columbia's Kwikwetlem First Nation (population: 82), earned $914 291 (Geddes 2014).

Gender is the final dimension of Indigenous diversity that I will mention. In some respects, Indigenous women are worse off than Indigenous men. Their average annual income is about 84 percent of the average annual income of Indigenous men, and their unemployment rate is about 2.5 percentage points higher. The incidence of domestic abuse among Indigenous women is 2.5 times higher than it is among non-Indigenous women. Remarkably, nearly 38 percent of murdered women in Canada in recent years have been Indigenous (O'Donnell and Wallace 2011: 40, 42). As is the case for all other ethnic and racial groups, most of these murders were committed by family members or intimate partners (Galloway 2015). However, a higher proportion of murders of Indigenous murders were committed by mere acquaintances or strangers, the latter mainly of non-Indigenous origin.

Notwithstanding such gender disparities, women stand at the forefront of some encouraging developments in the Indigenous community. Take 20-something Krystal Abotossaway, who is a member of Aundeck Omni-Kanging First Nations and Chippewas of Rama First Nations. She holds a degree in human resources management from York University in Toronto. Her job is to identify and hire talented young Indigenous workers for the Royal Bank of Canada. Her goal, she says, is to help ensure that Indigenous Canadians become important decision makers throughout the private sector (Grant 2015).

While there were just a few hundred Indigenous students in Canada's colleges and universities in the 1970s, there are now more than 30 000. Two-thirds of them are women (Friesen 2013). Research by a TD Bank economist found that Indigenous women have experienced a larger bounce-back in employment since the 2008–09 recession than either Indigenous men or non-Indigenous Canadians. Moreover, their employment growth has been especially rapid in the high-paying knowledge sector: finance, education, and professional services, including jobs in accounting and the law. Consequently, the wage gap between Indigenous and non-Indigenous women is narrowing (DePratto 2015). "They're leading the way," according to the economist who authored the report (quoted in Grant 2015).

Women have traditionally played an important leadership role in many First Nations bands. Now that they are gaining postsecondary education in substantial numbers, it is perhaps not surprising to discover that they are also playing a major role in important social movements that demand increased recognition of Indigenous people's rights. For example, the Idle No More movement erupted in November 2012 to promote land claims, environmental protection, and other issues of concern to Canada's Indigenous people. It was founded and led by four Saskatchewan First Nations women. Over the next 14 months, the movement organized teach-ins, rallies, flash mobs, marches, round dances, and blockades across the country. Idle No More and the similar social movements that preceded it are among the main forces that have compelled the Canadian public and government to recognize Indigenous rights, resulting in gradual improvements in the community's quality of life (Wotherspoon and Hansen 2013).

LAND AND PROTEST

Indigenous organizations—the Inuit Tapiriit Kanatami, the Métis National Council, the Assembly of First Nations (representing Status Indians), and the Congress of Aboriginal Peoples (representing mainly non-Status Indians)—aim to induce the Canadian government to honour its perceived treaty and moral obligations.

Treaties are legal agreements between the Crown and Indigenous people to exchange certain rights. Governments interpret treaties narrowly as real estate deals in which Indigenous groups sell their interest in large parcels of land in exchange for reserves and modest one-time and ongoing payments. Governments base their interpretation on a strict reading of treaty texts. In contrast, Indigenous groups interpret treaties broadly. They believe treaties establish relationships between autonomous nations that agreed to share Canada's land and resources for the mutual benefit of both parties.

Indigenous leaders base their interpretation on what was said during negotiations, often in Indigenous languages, and finalized ceremonially. The basis for their interpretation stands to reason given that most treaties were signed in the 1700s and 1800s, when the overwhelming majority of Indigenous people could not read or write. Over the generations, elders have been trained in oral history to become authorities on the intent of treaties. It was only in 1990 that the Supreme Court of Canada ruled that treaties should be interpreted liberally and uncertainties resolved in favour of Indigenous people. Still, most treaties continue to be in dispute and are the subject of ongoing litigation (Asch 2014; Hall and Albers 2015 [2011]).

Sometimes disputes boil over into Indigenous protest. Occasionally protest becomes violent. For example, in 1990, a disagreement between the Mohawk of Kanesatake in Quebec and the town council of Oka resulted in a 78-day standoff involving protesters, police, and the army over a golf course and condominiums that non-Indigenous developers planned to build on disputed land that included a Mohawk burial ground. Indigenous groups from across the country rallied to support the Mohawk. After protesters blocked access to the area, the Sûreté du Québec used tear gas and concussion grenades to break the blockade. One police officer was killed in the assault and a Mohawk elder died of a heart attack. The Sûreté du Québec then blockaded roads leading to the Kanesatake reserve. The Mohawk responded by halting all traffic on the Mercier bridge, cutting off access between the southern suburbs of Montreal and the Island of Montreal. As tempers flared, the RCMP and 2500 army troops were brought in to make arrests. Although that brought an end to the violence, the dispute ended only when plans for the golf course and condominiums were cancelled (Marshall 2014 [2013]).

Dalhousie University sociologist Howard Ramos documented the frequency of Indigenous protests between 1950 and 2000 (Ramos 2006, 2008). He found that they tended to occur in waves, with peaks occurring in 1969, 1974, 1981–82, 1989–90, 1995, and 1999. Ramos also discovered that protest frequency increased as Indigenous people formed new political organizations, received infusions of federal funds for organizational purposes, became the subject of increased media attention, and won land claims in the courts. In other words, Indigenous power has grown with increased access to political resources, growing public awareness of Indigenous issues, and encouragement resulting from successful court battles.

WHAT DO INDIGENOUS PEOPLE WANT?

Novelist Thomas King provides a terse but accurate answer to this question: Indigenous people form a socially heterogeneous group with a wide variety of interests and desires. They all want a future of their own making, but that future looks different depending on whom you ask (King 2012: 193, 215). Some, like the Royal Bank human resources manager quoted earlier, want postsecondary education and decision-making power in the private sector. Others want to return to the land and to traditional hunting, fishing, and other food-acquiring practices (Samson, Wilson, and Mazower 1999: 25). Indigenous people are much like everyone else, so there is no end to the variety of things they want.

One thing is clear, however: Regardless of what the future looks like to any particular individual or subgroup, opportunities for Indigenous people to create a future of their own making need to be based on two principles. First, land claims will have to be resolved in line with the 1990 Supreme Court ruling that treaties should be interpreted with a bias in favour of Indigenous people. Second, Indigenous people will require increased autonomy in running their own affairs. The opposite approach—forcing people off their land, ruining their food supplies, demanding their assimilation, making most of them dependent on the federal government—has produced one life-ruining disaster after another. In contrast, the favourable resolution of land claims will give Indigenous people control over some of the natural resources that were taken from them, while increased autonomy will allow them to use those resources as they see fit.

The downside of such a program is that it requires non-Indigenous Canadians to recognize that, for half a millennium, they and their ancestors have taken pretty much what they wanted for their own benefit, everything from land for golf courses and major city downtowns to water for hydroelectric power generation and oil sands production. Recognition that such resources should be used for the mutual benefit of Indigenous and non-Indigenous Canadians would require a substantial redistribution of wealth-generating resources. And few people give up their advantages without a fight or least a set of rationalizations for resisting change.

For example, many white Canadians point out that their parents or ancestors arrived in Canada just 25 or 50 or 100 years ago with little or nothing. They worked hard so their offspring could attend university and become professionals with good incomes. What happened centuries ago to Indigenous Canadians, they continue, is not their fault and therefore not their problem.

Surely such people have every right to be proud of their achievements and those of their family members. They *should* be grateful to the country that gave them abundant opportunities to succeed in life. But, just as surely, they should recognize that their family members took advantage of a hierarchical social structure built on colonialism and racial discrimination that was already in place when they arrived. The house they grew up in may have been constructed on land taken by force or by guile from the people who lived on it first and considered it theirs for millennia. The job opportunities that were open to them were based on the near-total destruction of the country's Indigenous people. In that sense, all white people are indebted to Indigenous Canadians. If non-Indigenous Canadians are to help improve the plight of Indigenous Canadians, we need to mix our pride with equal measures of humility and responsibility.

CRITICAL THINKING EXERCISES

1. Why have the Canadian public and the federal government been reluctant to recognize the full range of rights that Indigenous people think they deserve? Why do you agree or disagree with the reasoning of the Canadian public and the federal government?
2. What are the most effective ways for Indigenous people to influence the Canadian public and the federal government to recognize their rights—voting, engaging in legal proceedings, petitioning, organizing demonstrations and marches, or using violence? What are the advantages and disadvantages of each of these approaches?

APPENDIX

Russell Moses (1932–2013) was a member of the Delaware band of the Six Nations of the Grand River. He served in the Royal Canadian Navy and the Royal Canadian Air Force. After military service, he joined the Indian Affairs Branch of the Department of Citizenship and Immigration, eventually becoming special assistant to Jean Chrétien, Minister of Indian Affairs and Northern Development (Moses 2013).

When he was eight years old, Russell Moses was placed in a residential school outside Brantford, Ontario. In 1965, an official from the Indian Affairs Branch of the Department of Citizenship and Immigration asked Mr. Moses to describe his residential school experience. This is his reply (a copy of the original letter can be found online at nelson.com/student):

MOHAWK INSTITUTE – 1942–47

First, a bit of what it was like in the "good old days".

In August 1942, shortly before my 9th birthday a series of unfortunate family circumstances made it necessary that I along with my 7 year old sister and an older brother, be placed in the Mohawk Institute at Brantford, Ontario.

Our home life prior to going to the "Mohawk" was considerably better than many of the other Indian children who were to be my friends in the following five years. At the "mushole" (this was the name applied to the school by the Indians for many years) I found to my surprise that one of the main tasks for a new arrival was to engage in physical combat with a series of opponents, this was done by the students, so that you knew exactly where you stood in the social structure that existed.

The food at the Institute was disgraceful. The normal diet was as follows:

Breakfast – two slices of bread with either jam or honey as the dressing, oatmeal with worms or corn meal porridge which was minimal in quantity and appalling in quality. The beverage consisted of skim milk and when one stops to consider that we were milking from twenty to thirty head of pure bred Aolstain cattle, it seems odd that we did not ever receive whole milk and in my five years at the Institute we never received butter once.

This is very strange, for on entering the Institute our ration books for sugar and butter were turned in to the management – we never received sugar other than Christmas morning when we had a yearly feast of one shreaded wheat with a sprinkling of brown sugar.

Lunch – At the Institute this consisted of water as the beverage, if you were a senior boy or girl you received (Grade V or above) one and half slices of dry bread and the main course consisted of "rotten soup" (local terminology) (i.e. scraps of beef, vegetables some in a state of decay.) Desert would be restricted to nothing on some days and a type of tapioca pudding (fish eyes) or a crudely prepared custard, the taste of which I can taste to this day. Children under Grade V level received one slice of dry bread – incidentally we were not weight watchers.

Supper – This consisted of two slices of bread and jam, fried potatoes, NO MEAT, a bun baked by the girls (common terminology – "horse buns") and every other night a piece of cake or possibly an apple in the summer months.

The manner in which the food was prepared did not encourage over-eating. The diet remained constant, hunger was never absent. I would say here that 90% of the children were suffering from diet deficiency and this was evident by the number of boils, warts and general malaise that existed within the school population.

I have seen Indian children eating from the swill barrel, picking out soggy bits of food that was intended for the pigs.

At the "mushole" we had several hundred laying hens (white leghorn). We received a yearly ration of one egg a piece – this was on Easter Sunday morning, the Easter Bunny apparently influenced this.

The whole milk was separated in the barn and the cream was then sold to a local dairy firm, "The Mohawk Creamery", which I believe is still in business. All eggs were sold as well as the chickens at the end of their laying life – we never had chicken – except on several occasions when we stole one or two and roasted them on a well concealed fire in the bush – half raw chicken is not too bad eating!

The policy of the Mohawk Institute was that both girls and boys would attend school for half days and work the other half. This was Monday to Friday inclusive. No school on Saturday but generally we worked,

The normal work method was that the children under Grade V level worked in the market garden in which every type of vegetable was grown and in the main sold – the only vegetables which were stored for our use were potatoes, beans, turnips of the animal fodder variety. The work was supervised by white people who were employed by the Institute and beatings were administered at the slightest pretext. We were not treated as human beings – we were the Indian who had to become shining examples of Anglican Christianity.

I have seen Indian children having their faces rubbed in human excrement, this was done by a gentleman who has now gone to his just reward.

The normal punishment for bed wetters (usually one of the smaller boys) was to have his face rubbed in his own urine.

The senior boys worked on the farm – and I mean worked, we were underfed, ill clad and out in all types of weather – there is certainly something to be said for Indian stamina. At harvest times, such as potatoe harvest, corn harvest for cattle fodder – we older boys would at times not attend school until well on into fall as we were needed to help with the harvest.

We arose at 6:00 a.m. each morning and went to the barn to do "chores". This included milking the cattle, feeding and then using curry comb and brush to keep them in good mental and physical condition.

After our usual sumptuous breakfast we returned to the barn to do "second chores" 8:00 to 9:00 a.m. – this included cleaning the stables, watering the young stock and getting hay down out of mow, as well as carrying encilage from the silo to the main barn.

We also had some forty to eighty pigs depending on time of year – we never received pork or bacon of any kind except at Christmas when a single slice of pork along with mashed potatoes and gravy made up our Christmas dinner. A few rock candies along with an orange and Christmas pudding which was referred to as "dog shit" made up our Christmas celebrations. The I.O.D.E. sent us books as gifts.

Religion was pumped into us at a fast rate, chapel every evening, church on Sundays (twice). For some years after leaving the Institute, I was under the impression that my tribal affiliation was "Anglican" rather than Delaware.

Our formal education was sadly neglected, when a child is tired, hungry, lice infested and treated as a sub-human, how in heavens name do you expect to make a decent citizen out of him or her, when the formal school curriculum is the most disregarded aspect of his whole background. I speak of lice, this was an accepted part of "being Indian" at the Mohawk – heads were shaved in late spring. We had no tooth brushes, no underwear was issued in the summer, no socks in the summer. Our clothing was a disgrace to this country. Our so called "Sunday clothes" were cut down first world war army uniforms. Cold showers were provided summer and winter in which we were herded en masse by some of the bigger boys and if you did not keep under the shower you would be struck with a brass studded belt.

The soap for perfuming our ablutions was the green liquid variety which would just about take the hide off you.

Bullying by larger boys was terrible, younger boys were "slaves" to these fellows and were required to act as such – there were also cases of homosexual contact, but this is not strange when you consider that the boys were not even allowed to talk to the girls – even their own sisters, except for

15 minutes once a month when you met each other in the "visiting room" and you then spoke in hushed tones.

Any mail coming to any student or mail being sent was opened and read before ever getting to the addressee or to the Indian child – money was removed and held in "trust" for the child.

It was our practise at the "Mohawk" to go begging at various homes throughout Brantford. There were certain homes that we knew that the people were good to us, we would rap on the door and our question was: "Anything extra", whereupon if we were lucky, we would be rewarded with scraps from the household – survival of the fittest.

Many children tried to run away from the Institute and nearly all were caught and brought back to face the music – we had a form of running the gauntlet in which the offender had to go through the line, that is on his hand and knees, through widespread legs of all the boys and he would be struck with anything that was at hand – all this done under the fatherly supervision of the boys' master. I have seen boys after going through a line of fifty to seventy boys lay crying in the most abject human misery and pain with not a soul to care – the dignity of man!!

As I sit writing this paper, things that have been dormant in my mind for years come to the fore – we will sing Hymn No. 128!!

This situation divides the shame amongst the Churches, the Indian Affairs Branch and the Canadian public.

I could write on and on – and some day I will tell of how things used to be – sadness, pain and misery were my legacy as an Indian.

The staff at the Mohawk lived very well, separate dining room where they were waited on by our Indian girls – the food I am told, was excellent.

When I was asked to do this paper I had some misgivings, for if I were to be honest, I must tell of things as they were and really this is not my story, but yours.

There were and are some decent honourable people employed by the residential schools, but they were not sufficient in number to change things.

SUGGESTED IMPROVEMENTS FOR RESIDENTIAL SCHOOLS

1. Religion should not be the basic curriculum, therefore, it is my feeling that non-denominational residential schools should be established. (dreamer)
2. More people of Indian ancestry should be encouraged to work in residential schools as they have a much better understanding of the Indian "personality" and would also be more apt to be trusted and respected by the students.
3. Indian residential schools should be Integrated – the residential school should be a "home" rather than an Institute.

4. Salaries paid to the staff members should be on a par with industrie – otherwise you tend to attract only social misfits and religious zealots.
5. The Indian students should have a certain amount of work (physical) to do – overwork is no good and no work is even worse. I believe that a limited amount of work gives responsibility to the individual and helps him or her to develop a well-balanced personality.
6. Parents of Indian children should be made to contribute to the financial upkeep of their children – I realise that this would be difficult, but it at least bears looking into.
7. Each child should be given individual attention – get to know him or her – encourage leadership, this could be accomplished by giving awards for certain achievements.
8. Last, but most important, solicit ideas from the students, we adults do not know all the answers.

SUMMATION – The years that an Indian child spends in an Indian residential school has a very great deal to do with his or her future outlook on life and in my own case it showed me that Indian are "different", simply because you made us different and so gentlemen I say to you, take pains in molding, not the Indian of to-morrow, but the Canadian citizen of to-morrow. FOR "As ye sow, so shall ye reap".

Russell Moses

Source: Reprinted with permission from John Moses.

Aaron Millard

8 Sociology as a Vocation

IT'S ALL ABOUT THE CONTEXT

If this book has any value, then the next time you hear about people dying, you won't just ask, "What did they die of?" Instead, like a sociologist, you'll also ask, "What was the social context of their death?" Analytically and practically, a world of difference separates the two questions, as you learned in the preceding chapters.

Specifically, by investigating six social contexts in detail—the American inner city, the world of Palestinian suicide bombers, the hurricane-prone Caribbean Basin and the coast of the Gulf of Mexico, the environmental causes of cancer, the gender risks associated with womanhood, and the history of Indigenous–settler relations in Canada—we learned much of practical importance for improving the quality of life and increasing life expectancy.

We learned, first, that the defiant cry for identity known as hip hop has unfortunate consequences. Hip hop encourages deadly violence among some people. It also diverts attention from mobility strategies that are more likely than gangsterism to meet with success and thus improve the quality of life.

Second, we learned, that the policy of retaliation that is often involved in suicide bombings and targeted killings deepens the resolve of both sides in the Israeli–Palestinian conflict to kill each other. It thus moves both parties further away from a negotiated peace.

Our third lesson was that the lack of disaster planning in some hurricane-prone regions greatly increases the chance that the most vulnerable members of society will perish.

We learned, fourth, that, although it may not be possible to eliminate all cancers, we can rid ourselves of many if not most of them by removing their known causes from our environment.

Fifth, we learned that the level of gender risk that women face, while influenced by a range of powerful social forces, can be lowered by means of human intervention.

And our sixth lesson was that blaming Indigenous Canadians for their troubled history and their present state of affairs obscures the role of settlers and their offspring in perpetrating what I have advisedly termed genocide.

These lessons are sufficiently instructive that I feel wholly justified in claiming that sociology is a life or death issue.

But *caveat emptor*—let the buyer beware. I wouldn't feel comfortable selling you a bill of goods without telling you exactly what the purchase involves. Let me therefore close by offering a thumbnail sketch of how sociologists go about their business. That way, you can make an informed choice about how heavily, if at all, you want to buy into the discipline.

I can best describe what sociologists do by focusing on four terms: *values*, *theory*, *research*, and *social policy*. As you will now learn, sociologists' ideas about what is good and bad (their **values**) lead them to choose certain subjects for analysis and help them formulate tentative, testable explanations of the social phenomena that interest them (**theories**). Sociologists then systematically collect and analyze data to see if their tentative explanations are valid (**research**). Finally, based on the results of their research, they propose rules and regulations to govern the actions of organizations and governments in the hope of correcting social problems (**social policies**). Individual sociologists may specialize in theory construction, empirical research, or policy analysis, but the sociological enterprise as a whole involves all of these activities.

DOING SOCIOLOGY

Values

> There never has been, nor can there be, any human search for truth without human emotions.
>
> —*Vladimir Lenin (1964: 260)*

Someone once told Émile Durkheim that the facts contradicted one of his theories. "The facts," Durkheim replied, "are wrong" (quoted in Lukes 1973: 52). Likewise, when a colleague's experimental evidence appeared to challenge his special theory of relativity, Albert Einstein reacted in essence with, "so much the worse for the facts" (Clark 1971: 144).

What's wrong with this picture? Aren't scientists' opinions supposed to be governed by respect for evidence? How then could a founding father of sociology and the greatest physicist of the 20th century react in such a cavalier manner to facts that apparently falsified their views?

Actually, there is nothing wrong with this picture. Great scientists are human beings, and they can be as pigheaded as you and me. Having invested so much time and energy developing pet theories, they may cling to them stubbornly even if evidence suggests that they are wrong. What is intriguing is that, despite such common human failings as obstinacy, vanity, short-sightedness, and narrow-mindedness, scientists routinely arrive at knowledge that is widely considered valid.

Scientists are able to do so because people have designed the *institution* of science to eliminate ideas that can't be supported by evidence. The norms of science require that scientists publicize their findings and their methods of discovery, and that other scientists (with their own pet theories, prejudices, and professional rivalries) carefully scrutinize the work of their colleagues. Science is not quite a blood sport, but it does stimulate a lot of conflict over the validity of ideas. It is therefore not recommended for the faint of heart. Out of this conflict there typically emerges a temporary consensus about what constitutes valid knowledge. A higher level of consensus exists in physics than in sociology, but in all disciplines, the community of scholars playing by the rules of the scientific game determines the prevailing consensus.[1] You don't get special consideration if you're a Durkheim or an Einstein. Consensus, not authority, rules.

I would not, however, want to leave you with the impression that scientists' values are always a danger to science. To the contrary, values are critically important in the scientific enterprise. They first play a role in determining what we want to study. As Max Weber wrote, we choose to study "only those segments of reality which have become significant to

[1] Because the rewards for innovation are many and scientists are only human, some of them are tempted to break the rules. They may falsify data, for example. But unlike ordinary criminals, who are sometimes able to conceal their illicit activities, scientists work in the open. As a result, the scrutiny of the scientific community almost always ensures that they get caught, usually sooner rather than later.

us because of their value-relevance" (Weber 1964). Values also fire our imagination and our intuition. They shape our ideas about how the parts of society fit together, what the ideal society should look like, which actions and policies are needed to help us reach that ideal, and which theories offer the best explanations for the phenomena that interest us (Edel 1965).

True, when values infuse science, they may lead to subjectivity or bias of one sort or another. But that matters little as long as scientists allow their peers to judge the worth of their ideas and form a binding consensus—and as long as outside forces such as governments and corporations don't interfere with their ability to do so. The scientific community's consensus is precisely what is meant by objectivity, and as long as consensus is allowed to crystallize and prevail, bias will be minimized. Values are thus an invaluable source of scientific creativity. They become problematic only if the scientific community fails to remain vigilant in its scrutiny of theory construction, empirical research, and policy proposals.

Theory and Research

> Practice should always be based upon a sound knowledge of theory. . . .
> [W]ithout it nothing can be done well. . . .
>
> —*Leonardo da Vinci (1956: 910)*

As noted earlier, sociological theories are tentative, testable explanations of some aspect of social life. They state how and why certain social facts are related. For example, in his theory of suicide, Durkheim related facts about suicide rates to facts about social solidarity. Doing so allowed him to explain suicide as a function of social solidarity (see Chapter 1).

Non-testable explanations aren't scientifically useful. Some people believe that wealthy, unknown conspirators manipulate the price of gold and benefit from the price swings. Others believe that their physical ailments are caused by space aliens who routinely abduct them. Such theories are unscientific because they are not testable. Nobody has so far been able to observe, count, analyze, and interpret data that would allow us to decide whether they are valid. Note also that because scientific theories are testable, they are only tentative explanations. Research may disprove them at any time.

Although a theory may be little more than a hunch drawn from the experience of everyday life (Einstein 1954: 270), it is a start. Without some conjecture as to how facts are related, it is impossible to conduct meaningful research. That is what Leonardo da Vinci meant when he said that people cannot do anything well if they engage in practice (research, and action based on that research) without theory. Starting research before developing a clear theoretical statement of how the relevant social facts might be related is a recipe for wasting years collecting and analyzing data in the field, in the archives, or at a computer. The reverse is also problematic. Theorizing without research is about as useful to science as exercise without movement is useful to the human body. From the comfort of an armchair, one can entertain the wildest of conjectures. The reality check known as research brings theorists down to earth and turns speculation into knowledge.

A stylized version of the sociological research process is presented in Figure 8.1.[2] As you already know, research can begin only when sociologists identify topics they find significant (step 1) and hypothesize how observable facts central to their topic are related (step 2). Their next task is to review the existing theoretical and research literature (step 3). Reading previous work on the topic of interest stimulates one's sociological imagination and refines one's thinking. Besides, there is no sense rediscovering what earlier researchers have already found. Reviewing the relevant theoretical and research literature is an indispensable means of avoiding duplication of effort and learning from past insights, discoveries, and mistakes.

Figure 8.1 The Research Process

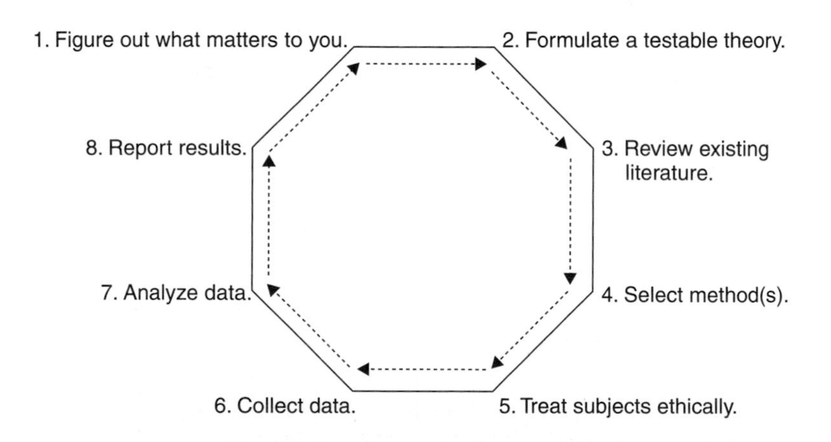

[2] I say "stylized" because research is typically less orderly than Figure 8.1 makes it appear. For example, researchers usually jump back and forth between stage 2 (formulate a testable theory) and stage 7 (analyze data) many times before proceeding to stage 8 (report results).

Different kinds of research questions require different research methods (step 4; Babbie and Roberts 2017). In general, sociologists adopt more precise methods when they research topics that are more theoretically refined. Consider surveys, the most widely used research method in sociology. Surveys involve asking a representative sample of people drawn from the population of interest a set of standardized questions about their attitudes, knowledge, and/or behaviour. Surveys may be conducted face to face, over the phone, through the mail, or via the Internet.

For the most part, the people who respond to surveys (**respondents**) are presented with a list of allowable answers from which to choose. But insofar as surveys require that researchers know in advance what questions to ask and the range of possible answers to their questions, surveys are not always an appropriate research tool. Particularly in areas where theory is not highly developed—in **exploratory research**—it usually makes sense to conduct interviews that are less structured and more open-ended. The interviews don't define a range of allowable responses and they allow the spontaneous introduction of new questions whose importance emerges only in the course of the interview.

In many cases, even conducting formal interviews is premature. For research where theory is poorly developed and researchers have only a hunch about how observable facts may be related, one or another form of field research is often the method of choice. For example, **participant observation** involves spending considerable time in the social setting of the people who interest the researchers. It allows researchers not just to step back and observe their subjects' milieu from an outsider's point of view, but also to step in and develop an understanding of the meaning people attach to the elements that populate their social world. From such observations, researchers may develop theoretical formulations that will permit more refined testing of ideas by means of structured interviews and surveys.

Many sociologists regard experiments as too artificial to provide valid knowledge about most forms of social behaviour because they remove people from their social settings (recall our discussion of experiments on the effects of media violence in Chapter 2). As a result, the use of experiments in sociology is not widespread. In contrast, various methods that use historical records and existing documents as data are increasingly popular. For instance, sociologists have interpreted published histories to develop important theories about the uneven spread of industrial capitalism around the world and the turn of some countries to democracy, others to authoritarianism (Moore 1967; Wallerstein 1974–89).

Once they leave their offices and begin collecting data, researchers must keep in mind four ethical commandments (step 5):

1. *Respect your subjects' right to safety*. Do your subjects no harm and, in particular, give them the right to decide whether and how they can be studied.
2. *Respect your subjects' right to informed consent*. Tell subjects how the information they supply will be used and allow them to judge the degree of personal risk involved in supplying it.
3. *Respect your subjects' right to privacy*. Allow subjects the right to decide whether and how the information they supply may be revealed to the public.
4. *Respect your subjects' right to confidentiality*. Refrain from using information in a way that allows it to be traced to a particular subject.

Ethical considerations are not restricted to the data-collection phase of research (step 6). They come to the fore during the data analysis and reporting phases, too. There, two more commandments apply:

5. *Do not falsify data*. Report findings as they are, not as you would like them to be.
6. *Do not plagiarize*. Explicitly identify, credit, and reference authors when making use of their written work in any form, including Web postings.

If theories propose conjectures about the way certain social facts are related, data analysis involves determining the degree to which the information at one's disposal conforms to the conjectures (step 7). Information can often be usefully turned into numbers. For example, if the allowable responses to a survey question are "strongly disagree," "disagree," "neither disagree nor agree," "agree," and "strongly agree," the responses can be assigned the numbers 1 through 5, respectively. The advantage of turning information into numbers is that it permits statistical analysis using computers.

You can appreciate why computer-assisted statistical analysis is useful by imagining a spreadsheet with 240 000 cells. Since it is not unusual for surveys to ask 1200 respondents 200 questions each, data matrices of this size are common in sociology (1200 × 200 = 240 000). Each of the spreadsheet's 1200 (horizontal) rows is reserved for an individual respondent; each of its 200 (vertical) columns is reserved for responses to one question in the survey.

A vast amount of information is represented in that spreadsheet. Human beings don't have the ability to scan, unaided, a spreadsheet with nearly a quarter of a million cells and spot patterns in the numbers that would lead them to conclude that the data conform more or less closely to their theoretical conjectures. Computer-assisted statistical analysis is simply a means of finding patterns in the data matrix—patterns that allow the researcher to see whether conjectures hold up. In milliseconds, an ordinary personal computer equipped with the right software can add up a column of 1200 numbers corresponding to one opinion question in a survey and then divide the sum by the number of respondents who answered the question, telling us the average opinion of the respondents. Or it can compute a statistic that tells the researcher how responses to one opinion question vary with responses to another opinion question, so he or she can learn how, if at all, the opinions are related. More complex statistical analyses showing, say, the unique and combined effects of class, race, gender, religion, place of residence, and age on political preference might take a few more milliseconds. The point is that having such pattern-finding power at one's disposal increases the researcher's capacity to test, reformulate, and retest theories quickly and creatively.

The final stage in the research process involves publishing the results of one's analysis in a report, journal, or book (step 8). Doing so allows other sociologists to pore over the research so that errors can be corrected and better questions can be formulated for future research. It also allows people to apply research findings to social policy.

Some sociological research is conducted with particular policy aims in mind and therefore has a direct impact on the formulation of rules, regulations, laws, and programs by organizations and governments. Most sociological research, however, has only an indirect impact on social policy. For example, no organization or government hired me to write this book, and I entertain no illusion that officials will read it and suddenly decide they need to change the way they do business. Like many sociologists, I aim to influence the educated public and, in particular, young adults. I hope that my work will contribute, however slightly, to enlighten people about the opportunities and constraints that help shape their thoughts and actions, and assist them in making personal, organizational, and political decisions that will allow them and their fellow citizens to live happier and longer lives than would otherwise be possible. It's an immodest hope but, I think, not entirely unrealistic.

CAREERS

What about your hopes? Might they include the study of sociology? To make an informed decision on such an important issue, you need information on what you can expect from a sociology degree. In this connection it is instructive to examine the careers pursued by sociology graduates. About 1.35 million Americans and 150 000 Canadians graduate with a bachelor's degree every year. Roughly 29 000 of them major in sociology. About 2110 will eventually receive a master's degree in sociology and 657 will earn a Ph.D. degree in sociology (estimated from National Center for Education Statistics 2004, 2005).[3] What do these graduates do with their degrees?

Many people with a *graduate* degree in sociology teach and conduct research in colleges and universities. Research is a bigger part of the job in more prestigious institutions.

Many sociologists with a graduate degree conduct research and give policy advice in institutions outside the system of higher education. The number of research- and policy-related jobs for sociologists is growing faster than the number of teaching jobs. In government, sociologists help formulate research-based policy in such areas as health, social welfare, economic and social development, the elderly, youth, criminal justice, science, and housing (House 2005). Nongovernmental agencies that employ sociologists include professional and public-interest associations and trade unions. In the private sector, sociologists practise their craft in firms that specialize in public opinion polling, management consulting, market research, standardized testing, and evaluation research, which assesses the impact of policies and programs.

[3] These estimates are based on U.S. data for graduates in the academic year 2002–03. I added 10 percent to the U.S. figures to take account of Canada's population and its slightly lower graduation rate. I also assumed that the ratio of master's to bachelor's degree graduates and the ratio of doctoral to bachelor's degree graduates in 2002–03 were equal to the proportion of bachelor's students who eventually graduate with master's and Ph.D. degrees.

Although comparable Canadian data are unavailable, in the United States in 2015, average annual earnings stood at US$71 220 for all life, physical, and social science occupations, $78 520 for all social science occupations separately, and $82 100 for sociologists only (U.S. Department of Labor 2015). Canadian research comparing the earnings of majors in the social sciences and humanities shows that graduates in no field earn significantly more than sociology majors do (Davies and Walters 2011).

An *undergraduate* major in sociology is excellent preparation for many fields other than sociology (Stephens 1998). American research shows that about 40 percent of people who major in sociology intend to go to graduate or professional school. Of those who aspire to do so, more than 82 percent expect to choose fields other than sociology, including education, social work, law, and criminology (Spalter-Roth, Erskine, Polsiak, and Panzarella 2005: 24–25). See Table 8.1 for a list of jobs most commonly held by Canadians with a B.A. in sociology.

From the point of view of the job market, the value of a sociology major lies partly in the generic skills it offers rather than any specifically technical skills associated with it (Davies and Walters 2011). Sociology majors (as well as majors in the other social sciences) tend to be adept in interpersonal relations, communication, decision making, critical thinking, conflict management, the assumption of authority, and the ability to apply abstract ideas to real-life situations. As a result, they can be trained quickly and effectively to do any number of jobs in the quickly growing service sector of the economy. Most employers know that so employment opportunities for social science graduates are generally good (Allen 1999).

POSITIVE AND NEGATIVE FREEDOM

Now that I have sketched what sociologists do and how you might fit into the sociological enterprise, there remains the task of summarizing the values, theories, and policies that I have introduced in this book.

Underlying my analyses is a particular conception of human freedom that I have come to value highly and that I must first clarify. Political philosophers distinguish negative freedom from positive freedom (Berlin 2002b; MacCallum 1967). **Negative freedom** is freedom *from* constraints that would otherwise prevent me from doing as

Table 8.1 Jobs Most Commonly Held by Canadians with BA in Sociology

Government
- affirmative action work
- community affairs
- development aide
- foreign service work
- human rights officer
- information officer
- legislative assistant
- personnel coordinator
- policy research
- urban/regional planner

Teaching / Education
- admissions counsellor
- alumni relations
- continuing studies
- post-secondary recruitment
- public health educator
- records and registration
- school counselling
- student development
- teacher

Social Research
- census officer/analyst
- consumer researcher
- data analyst
- demographer
- market researcher
- social research specialist
- survey researcher
- systems analyst

Business
- actuary
- administrative assistant
- advertising officer
- computer analyst
- consumer relations
- data entry manager
- human resources specialist
- insurance agent
- journalist
- labour relations officer
- market analyst
- merchandiser/purchaser
- personnel officer
- production manager
- project manager
- public relations officer
- publishing officer
- quality control manager
- real estate agent
- sales manager
- sales representative
- technical writing

Community Affairs
- addictions. counselling
- adoption counselling
- caseworker
- child development
- community organizer
- environmental organizer
- family planning
- fundraising
- gerontologist
- group home programmer
- health outreach work
- homeless/housing worker
- hospital administration
- housing coordinator
- marriage/family counselling
- occupational/career counsellor
- public health worker
- rehabilitation work
- residential planning
- social assistance advocate
- welfare counselling
- youth outreach

Source: Canadian Sociological Association and Department of Sociology, University of British Columbia. 2014. Opportunities in Sociology. *Montreal. Reprinted by permission of the Canadian Sociological Association.*

I wish. When I have the right to vote, express my opinions publicly, and associate with whomever I wish, I enjoy negative freedom. That is because I can engage in these activities only when nobody prevents me from doing so. Note that non-interference from the state is usually required if I am to enjoy negative freedom.

In contrast, **positive freedom** is the capacity *to* act rationally. It involves taking control of one's life and realizing one's best interests. When I receive a higher education and good medical treatment in the event of ill health, I enjoy positive freedom. That is because I require such benefits if I hope to realize my best interests. However, particular individuals may be incapable of understanding their best interests, so it is typically necessary for some higher authority, notably the state, to help define those interests and provide the benefits needed to attain them.

Some political philosophers argue that positive freedom and negative freedom are incompatible. If you have more positive freedom, they say, you must have less negative freedom, and vice versa. The notion that positive and negative freedoms are rivals was first proposed by Sir Isaiah Berlin in the mid-20th century (Berlin 2002b). Berlin argued that if the state tried to define the best interests of the citizenry and provide the benefits it deemed necessary to attain them, it could easily slip into authoritarianism. He pointed to the Soviet Union as a case in point. The rulers of the Soviet Union, guided by communist ideology, claimed to know what was required for the citizenry to realize its best interests. The state provided free education, child care, medical care, and so on, but people were not free to vote for the party of their choice, express their opinions publicly, or associate with whomever they chose.

In contrast to Berlin's argument, this book is based on the idea that positive and negative freedoms *are* generally compatible (Taylor 1985). Figure 8.2 supports the validity of my claim. It contains data on all 123 countries for which recent data on positive and negative freedom exist. It plots an index of negative freedom (along the horizontal axis) against an index of positive freedom (along the vertical axis). The index of negative freedom combines information on freedom to vote, freedom of the press, freedom of assembly, and other political and civil liberties in each country. It indicates how free people are from state oppression. The index of positive freedom combines information on average level of education, adult literacy, and life expectancy in each country. It indicates how free people are to realize their interests because they have the state support (through education and health care) to do so.

If Berlin's argument held, we would expect to discover an inverse relationship between positive and negative freedom: the more of one, the less of the other. But the trend line in Figure 8.2 shows just the opposite. The countries ranking highest on positive freedom also tend to rank highest on negative freedom. The countries ranking lowest on positive freedom also tend to rank lowest on negative freedom. Some countries deviate from this tendency, but the overall association between positive and negative freedom is strong ($r = 0.49$).

In general, then, and contrary to what Berlin argued, negative and positive freedoms are usually not rivals. In most cases, states that intervene to provide benefits deemed to be beneficial for the citizenry also *remove* political constraints from their citizens. States that constrain their citizens the most tend to provide *fewest* benefits to the citizenry. What gives the trend line in Figure 8.2 its upward slope are the mainly

Figure 8.2 Positive by Negative Freedom in 123 Countries

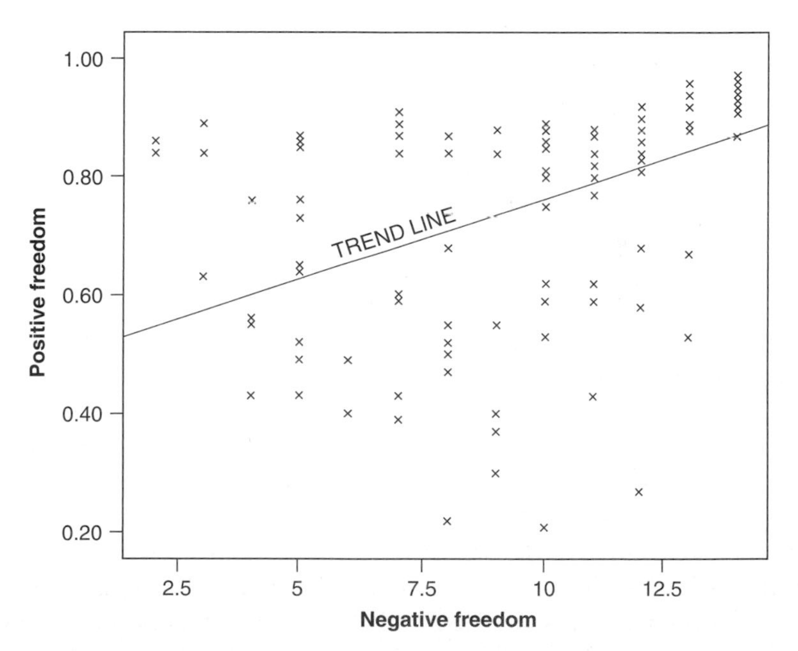

Note: Negative freedom is measured by the index of political and civil liberties, which is calculated annually by Freedom House, a non-partisan organization that promotes democracy around the world. Positive freedom is measured by life expectancy at birth, the literacy rate for people over the age of 14, and enrolment ratios at all levels of education, as calculated annually by United Nations' social scientists.

Sources: Compiled from Freedom House (2006); United Nations (2005: 219–22).

rich and moderately well-off countries in the upper right quadrant that have well-developed state services and are highly democratic, and the mainly poor and moderately well-off countries in the lower left quadrant that have less developed state services and authoritarian regimes.

PUTTING IT ALL TOGETHER

The idea that positive freedom is compatible with negative freedom runs through this book like a multicoloured thread. Consider the main arguments I made in each of the book's substantive chapters.

In the chapter on suicide bombers, my main theoretical argument boils down to the view that decreasing social solidarity among high-solidarity groups lowers the rate of what Durkheim called "altruistic suicide." With good reason, insurgent Palestinian groups believe that they and their people are threatened. Actions that reinforce that belief, such as targeted killings by the Israeli state, encourage some of them to take their own lives for the sake of their people. It follows that decreasing the perceived threat will lower social solidarity and the frequency of suicide attacks. I called for specific state actions to achieve that end: the elimination of targeted killings by Israel, and the use of powerful diplomatic and economic levers by the United States to compel both sides to negotiate a settlement that will recognize their territorial and security needs. The positive freedom to live in peace requires such brave and energetic state intervention in my view. I believe, moreover, that, absent their chronic conflict, Palestinians and Israelis would be able to remove restrictions on political and civil liberties that now constrain people in both societies. Peace would allow negative freedom to grow in the Middle East.

The main theoretical argument of the chapter on the myth of natural disasters is that when the state provides higher standards of security and well-being to the most vulnerable members of society, it increases their welfare and lowers their chance of death due to natural disasters. At the risk of raising a red flag, so to speak, I made the point by comparing the treatment of vulnerable populations in the United States and Cuba. I fully appreciate that some readers will have reacted negatively to my suggestion that the United States can learn a valuable lesson from a communist country. That is why I emphasized that Japan, too, takes exactly the measures that I think are necessary to protect vulnerable citizens without restricting anyone's negative freedom. The Japanese case demonstrates that freedom from excessive state authority and the freedom of society's most vulnerable members to live in relative safety are not incompatible.

The chapter on Indigenous Canadians demonstrates that when the state systematically neglects its most vulnerable members and acts in a way that denies them their right to life and collective identity, widespread death results. Privileged citizens may rationalize their actions by concocting stories about the inherent threat that the disadvantaged pose to security and to progress, and their actions may take centuries to exercise their full effect, but the stench of genocide by any other name smells just as sickening.

The stubborn smoker who insists that she ought to be able to live any way she wants is likely to oppose the policies I recommended in Chapter 5. If she studied Berlin's writings, she might say that if governments impose positive freedom by increasing taxes to fund wellness centres and adopt other cancer-prevention measures, the result must be more constraint on the citizenry and therefore less negative freedom. I claim she has things the wrong way around. Governments will take effective measures to prevent cancer only if an assertive citizenry demands it. More voting, more lobbying, more public expression of controversial ideas, more political mobilization—in short, more negative freedom—is necessary if government is to impose more positive freedom on the citizenry. Negative and positive freedoms are friends, not rivals.

The same holds true for women's safety. In the chapter devoted to that subject, I emphasized that the single most important determinant of danger to women worldwide is the willingness or reluctance of governments to implement policies fostering gender equality. In turn, an independent and strong women's movement (which, of course, includes sympathetic men) is required to push governments to act. Without the more vigorous exercise of negative freedom by a growing part of the citizenry, women cannot achieve the level of safety that men enjoy.

Finally, in the chapter on hip hop, we visited the American inner city, where the sense of community is weak and social institutions, especially families and schools, are crumbling. As Durkheim would lead us to expect in such an anomic social setting, the rate of suicide among inner-city black youth is high and rising quickly. Other forms of violence leading to death, notably homicide, also proliferate, as do cultural products such as hip hop that sometimes glorify violence.

I suggested that state intervention could help shore up and provide alternatives to faltering social institutions. More engaging and higher-quality schools, widely accessible child-care programs, and organized after-school athletic activities would, for example, make the inner city less anomic. Gun control would also help. Such policies would increase the positive freedom of residents in the inner city to live more secure lives and enjoy better prospects for upward mobility. They would also increase negative freedom by lowering the constraints on life imposed by coercive violence.

Such are my arguments. I do not ask you to agree with them. I do invite you to engage in a debate with your classmates and professors on the subjects I have raised, using logic, evidence, and the sociological perspective to add substance to your views. As you do so, you may discover, as I did when I was 19, that sociology is not just a course, but a calling.

CRITICAL THINKING EXERCISES

1. In one page, outline the roles of subjectivity and objectivity in scientific research.
2. In one page, explain the relationship between values, theory, research, and social policy.
3. In one page, explain the relationship between negative freedom and positive freedom.

Glossary

Absolute deprivation refers to long-standing poverty and unemployment.

The **age-standardized incidence rate** allows one to compare populations with different age structures in terms of how frequently a phenomenon occurs. Manipulating the data so the age distribution of one population is the same as the age distribution of another population ensures that any change we observe in incidence rates is not due to differences in the age structure.

Altruistic suicide, according to Durkheim, is suicide that occurs in high-solidarity settings.

Anomic suicide, according to Durkheim, is suicide that occurs when norms governing behaviour are vaguely defined.

The **cancer incidence rate** is the number of cases of cancer per 100 000 people in a population.

The **cancer paradox** is the fact that, although we know how to reduce the cancer incidence rate drastically, we have not been able to do so because (1) many people lack the resources to act, and (2) industry and government often resist vigorous preventative action.

A **cause-and-effect relationship** exists if (1) the cause precedes the effect, (2) the cause and effect are significantly correlated, (3) an observable mechanism links cause and effect, and (4) we rule out background factors that could plausibly be responsible for the correlation between cause and effect.

Citizenship refers to the rights of people to various protections under the law. To varying degrees, citizens of different countries have fought for and won civil rights (free speech, freedom of religious choice, justice under the law), political rights (freedom to vote and run for office), and social rights (freedom to receive a minimum level of economic security and participate fully in social life).

The **contact hypothesis** holds that prejudice will be reduced to the degree that majority and minority group members engage in friendly, informal, and personal interaction.

Culture is the sum total of shared and socially transmitted languages, beliefs, symbols, values, material objects, routine practices, and art forms that people create to help them survive and prosper.

Determinism is the belief that everything happens the way it does because it was destined to happen in just that way.

Egoistic suicide, according to Durkheim, is suicide that results from a lack of integration of the individual into society because of weak social ties to others.

Environmental racism is the tendency to heap environmental dangers on people who are disadvantaged and especially on members of disadvantaged racial minority groups.

Experiments are carefully controlled artificial situations that allow researchers to isolate presumed causes and measure their effects precisely. Typically, subjects are randomly divided into experimental and control groups. Only the experimental group is exposed to the presumed cause. The hypothesized effect is measured in the experimental and control groups before and after exposure. By comparing the measures of the effect before and after exposure, in both the experimental and control groups, experimenters can determine how much influence the presumed cause had on the hypothesized effect.

Exploratory research is conducted to formulate rather than test theory. Relatively unstructured and open-ended questions are usually asked in exploratory research.

Female-selective abortion involves terminating a pregnancy because the child is expected to be female.

Field research involves systematically observing people in their natural social settings.

First Nations are Status and non-Status Indians. See **non-Status Indians** and **Status Indians**.

Gender risk is the particular constellation of dangers associated with being a woman or a man.

Genocide is a denial of the right of existence of entire human groups. It is a crime under international law.

The **homicide rate** is the number of murders per 100 000 people in a population in a year.

The **infant mortality rate** is the number of babies who die before the age of one per 100 000 live births.

Intersectionality is the tendency for class and ethnicity/race to amplify the effect of gender on a range of behaviours, including violence.

The **Inuit** are Indigenous Canadians who reside in the country's northern regions.

The **least-squares regression line** (or trend line) is a straight line in a two-dimensional graph that is drawn so as to minimize the sum of the squared distances between each data point and the line itself.

Life expectancy is the average number of years a person can be expected to live.

Markets are social relationships that regulate the exchange of goods and services. In a market, the prices of goods and services are established by how plentiful they are (supply) and how much they are wanted (demand).

The **Métis** are Canadians of Indigenous and European (usually French) origin. They reside mainly in the western provinces and Ontario.

The **mortality rate** is the number of deaths per 100 000 people in a population in a year.

Negative freedom is the absence of constraint. It is freedom from obstacles that would otherwise prevent one from doing as one wishes.

Non-Status Indians are people who were once Status Indians but lost the title. See **Status Indian**.

Official statistics are numerical data compiled by state organizations for purposes other than sociological research.

Participant observation research involves spending considerable time in the natural social setting of the people who interest the researchers, allowing the researchers not just to step back and observe their subjects' milieu from an outsider's point of view, but also to step in and develop an understanding of the meaning people attach to the elements that populate their social world.

Patriarchy is the system of inequality between women and men that exists in most societies to varying degrees. It is characterized by male exclusivity in inheriting property, newly married couples establishing residence with or near the husband's parents, offspring assuming the surname of their father and tracing their lineage through the father's family, and the predominance of male authority inside and outside the family.

A **population** is the entire group that a researcher wants to learn about.

Positive freedom is the capacity to take control of one's life and realize one's best interests. It is freedom to act rationally.

Power is the ability to realize one's will, even against the resistance of others.

Races are categories of people defined not so much by biological differences as by social forces. That is to say, racial distinctions are typically made and reinforced by advantaged people for the purpose of creating and maintaining a system of inequality.

A **rate** lets one compare groups of different sizes. To calculate the rate at which an event occurs, divide the number of times an event occurs by the total number of people to whom the event could occur in principle. Then, calculate how many times it would occur in a population of standard size (say, 100 000).

Relative deprivation refers to the growth of an intolerable gap between what people expect and what they get out of life.

Research is the process of systematically collecting and analyzing data to test theories.

Respondents are people who reply to surveys.

A **sample** is the part of a group of interest that researchers study to learn about the group as a whole (the population).

The **sex ratio** is the number of men per 100 women in a population.

Social action is human behaviour that is meaningful in the sense that it takes into account the behaviour of others.

A **social class** is a position occupied by people in a hierarchy that is shaped by economic criteria, including wealth.

Social interaction is a dynamic sequence of social actions in which people (or groups) creatively react to each other.

Social policies are rules and regulations that organizations and governments establish to manage social problems.

Social solidarity refers to (1) the degree to which group members share beliefs and values, and (2) the intensity and frequency of their interaction.

Social structures are relatively stable patterns of social relations that constrain and create opportunities for the way we think and act.

Sociology is the systematic study of human action in social context.

Status Indians are Indigenous people who live throughout Canada and are registered as "Indians" under the Indian Act. Most of them belong to a band that signed a treaty with the Crown.

In a **survey**, randomly selected people are asked questions about their knowledge, attitudes, or behaviour. Researchers aim to study part of a group (a sample) to learn about the whole group of interest (the population).

Theories are tentative, testable explanations of phenomena.

Treaties are legal agreements between the Crown and Indigenous people to exchange certain rights.

Upward mobility refers to movement up a system of inequality.

Values are ideas about what is good and bad.

Victim blaming involves holding the targets of wrongful action responsible for the harm that befalls them.

Voluntarism is the belief that people alone control their destiny.

References

Abu-Odeh, Lama. 1996. "Crimes of Honour and the Construction of Gender in Arab Societies." Pp. 141–94 in *Feminism and Islam*, edited by Mai Yamani. New York: New York University Press.

Agenda Inc. 2005. "American Brandstand 2005." Retrieved March 27, 2006 (http://www.agendainc.com/brandstand05.pdf).

Akwagyiram, Alexis. 2009. "Hip-Hop Comes of Age." *BBCNews*, October 12. Retrieved July 19, 2010 (http://news.bbc.co.uk/2/hi/8286310.stm).

Al Jandaly, Bassma. 2009. "Working after Retirement in the UAE." *Gulfnews.com*, May 2. Retrieved May 9, 2013 (http://gulfnews.com/uaessentials/residents-guide/working/working-after-retirement-in-the-uae-1.440878).

al-Quds. 2000–2005. Jerusalem. [Arabic: *Jerusalem*.]

al-Quds al-'Arabi. 2000–2005. London. [Arabic: *Arab Jerusalem*.]

"al-Ra'is: 'Amaliyat Nitanya Juramat did Sha'buna.'" [Arabic: "The President: 'Netanya Operation Is a Crime against Our People.'"] 2005. *al-Quds*, July 12: 1A. Jerusalem.

Alberta Geological Survey. 2009. "Alberta Oil Sands." Retrieved June 26, 2010 (http://www.ags.gov.ab.ca/energy/oilsands/alberta_oil_sands.html).

Ali, Kecia. 2003. "Honor Killings, Illicit Sex, and Islamic Law." *MANA*. Retrieved April 20, 2013 (http://www.manavzw.be/dossiers/Gender/Honor_killings,_illicit_sex,_and_islamic_law).

Allen, Robert C. 1999. *Education and Technological Revolutions: The Role of the Social Sciences and the Humanities in the Knowledge-Based Economy*. Ottawa: Social Sciences and Humanities Research Council of Canada.

Altman, Lawrence K. 2000. "Genomic Chief Has High Hopes, and Great Fears, for Genetic Testing." *New York Times*, June 27. Retrieved June 23, 2010 (http://www.nytimes.com).

American Society of Plastic Surgeons. 2016. "New Statistics Reflect the Changing Face of Plastic Surgery." Retrieved December 15, 2016 (https://www.plasticsurgery.org/news/press-releases/new-statistics-reflect-the-changing-face-of-plastic-surgery).

Amnesty International. 2009. "No More Stolen Sisters: The Need for a Comprehensive Response to Discrimination and Violence against Indigenous Women in Canada." Retrieved May 31, 2013 (http://www.amnesty.org/en/library/asset/AMR20/012/2009/en/1943e1ef-1d45-4c42-a991-c6f0eea23a97/amr200122009en.pdf).

Anderson, Craig, and Brad J. Bushman. 2002. "The Effects of Media Violence on Society." *Science* 295: 2377–79.

Anderson, Michael. 2003. "Reading Violence in Boys' Writing." *LanguageArts* 80: 223–31.

Anderson, Siwan, and Debraj Ray. 2012. "The Age Distribution of Missing Women in India." *Economic and Political Weekly*, December 1, pp. 87–95.

Arian, Asher. 2001. "Israeli Public Opinion in the Wake of the 2000–2001 Intifada." *Strategic Assessment* 4. Retrieved May 15, 2005 (http://www.tau.ac.il/jcss/sa/v4n2p.Ari.html).

__________. 2002. "A Further Turn to the Right: Israeli Public Opinion on National Security–2002." *Strategic Assessment* 5. Retrieved May 15, 2005 (http://www.tau.ac.il/jcss/sa/v5n1p4Ari.html).

Armenian National Institute. 2016. "Frequently Asked Questions about the Armenian Genocide." Retrieved December 30, 2016 (http://www.armenian-genocide.org/genocidefaq.html).

Arokiasamy, Perianayagam, and Srinivas Goli. 2012. "Provisional Results of the 2011 Census of India."*International Journal of Social Economics* 39: 785–801.

Asch, Michael. 2014. *On Being Here to Stay: Treaties and Aboriginal Rights in Canada*. Toronto: University of Toronto Press.

Associated Press. 2005. "Lil' Kim Sentenced to a Year in Prison." *MSNBC.com*. July 6. Retrieved April 20, 2006 (http://www.msnbc.msn.com/id/8485039).

Atran, Scott. 2003. "Genesis of Suicide Terrorism." *Science* 299: 1534–39.

Aylsworth, Laura, and Frank Trovato. 2015 [2006]. "Demography of Indigenous People." *Canadian Encyclopedia*. Retrieved December 30, 2016 (http://www.thecanadianencyclopedia.ca/en/article/aboriginal-people-demography).

Babbie, Reginald and Lance Roberts. 2017. *Fundamentals of Social Research*, 4th Canadian Ed. Toronto: Nelson.

Baer, Robert. 2006. "The making of a suicide bomber." *The Sunday Times*. September 3. Retrieved August 9, 2016 (http://www.thesundaytimes.co.uk/sto/news/article167649.ece).

Baird, P.A. 1994. "The Role of Genetics in Population Health." Pp. 133–59 in *Why Are Some People Healthy and Others Not?*, edited by Robert G. Evans, Morris L. Barer, and Theodore R. Marmor. Hawthorne, NY: Aldine de Gruyter.

Bakx, Kyle. 2016. "Can oil and gas wealth improve First Nations conditions?" *Globe and Mail* October 15. Retrieved December 30, 2016 (http://www.cbc.ca/news/business/lng-fortmckay-wet-suwet-en-1.3803517).

Barlow, Maude, and Elizabeth May. 2000. *Frederick Street: Life and Death on Canada's Love Canal.* Toronto: HarperCollins.

Bauman, Zygmunt. 1991 [1989]. *Modernity and the Holocaust.* Ithaca NY: Cornell University Press.

Bayles, Martha. 1994. *Hole in Our Soul: The Loss of Beauty and Meaning in American Popular Music.* Chicago: University of Chicago Press.

Becker, Ernest. 1971. *The Birth and Death of Meaning: An Interdisciplinary Perspective on the Problem of Man.* New York: Free Press.

__________. 1973. *The Denial of Death*. New York: Free Press.

Bell, Sonya. 2013. "Status of Women Drops Long-Standing Raison d'Être, Performance Targets." *iPolitics.ca*, April 4. Retrieved May 1, 2013 (http://www.ipolitics.ca/2013/04/04/status-of-women-drops-long-standing-raison-detre-performance-targets).

Ben Salem, Lilia. 2010. "Women's Rights in the Middle East and North Africa 2010—Tunisia." *Freedom House*. Retrieved April 23, 2013 (http://www.refworld.org/cgi-bin/texis/vtx/rwmain?page=topic&tocid=4565c2252f&toid=4565c25f3d1&publisher=&type=&coi=TUN&docid=4b99011cc&skip=0).

Berger, Peter L., and Thomas Luckmann. 1966. *The Social Construction of Reality: A Treatise in the Sociology of Knowledge.* Garden City, NY: Doubleday.

Berlin, Isaiah. 2002a. "Historical Inevitability." Pp. 94–165 in *Liberty*, edited by Henry Hardy. Oxford: Oxford University Press.

_________. 2002b. "Two Concepts of Liberty." Pp. 166–217 in *Liberty*, edited by Henry Hardy. Oxford: Oxford University Press.

Berube, Alan, and Steven Raphael. 2005. "Access to Cars in New Orleans." The Brookings Institution. Retrieved May 30, 2006 (http://www.brookings.edu/metro/20050915_katrinacarstables .pdf).

BlackDemographics.com. 2016. "Black Educational Attainment by the Numbers." Retrieved December 30, 2016 (http://blackdemographics .com/education-2/education).

Block, Fred, Anna C. Korteweg, and Kerry Woodward. 2006. "The Compassion Gap in American Poverty Policy." *Contexts* 5: 14–20.

Blumenthal, Sidney. 2005. "No One Can Say They Didn't See It Coming." *Salon.com*, August 31. Retrieved May 31, 2006 (http:// dir.salon.com/story/opinion/blumenthal/2005/08/31/disaster _preparation/index_np.html).

Bourne, Joel. K. 2004. "Gone with the Water."*National Geographic Magazine*, October. Retrieved May 30, 2006 (http://magma .nationalgeographic.com/ngm/0410/feature5/?fs=www3 .nationalgeographic.com).

Boyce, Jillian, and Adam Cotter. 2013. "Homicide in Canada 2012." *Juristat* 33, 1. Retrieved August 8, 2016 (http://www.statcan.gc .ca/pub/85-002-x/2013001/article/11882-eng.htm).

Briggs, Jean L. 1970. *Never in Anger: Portrait of an Eskimo Family*. Cambridge, MA: Harvard University Press.

Brophy, James T., Margaret M. Keith, Kevin M. Gorey, Isaac Luginaah, Ethan Laukkanen, Deborah Hellyer, Abraham Reinhartz, Andrew Watterson, Hakam Abu-Zahra, Eleanor Maticka-Tyndale, Kenneth Schneider, Matthias Beck, and Michael Gilbertson. 2006. "Occupation and Breast Cancer: A Canadian Case-Control Study." *Annals of the New York Academy of Science* 1076: 765–77.

Brophy, James T., Margaret M. Keith, Kevin M. Gorey, Ethan Laukkanen, Isaac Luginaah, Hakam Abu-Zahra, Andrew Watterson, Deborah Hellyer, Abraham Reinhartz, and Robert Park. 2007. "Cancer and Construction: What Occupational Histories in a Canadian Community Reveal." *International Journal of Occupational and Environmental Health* 13: 32–38.

Browne, Kevin D., and Catherine Hamilton-Giachritsis. 2005. "The Influence of Violent Media on Children and Adolescents: A Public-Health Approach." *The Lancet* 365: 702–10.

Brunhuber, Kim. 2015. "Hurricane Katrina: Half of New Orleans still feels left out of recovery." *CBC News*, August 28. Retrieved August 8, 2016 (http://www.cbc.ca/news/world/hurricane-katrina-half-of-new-orleans-still-feels-left-out-of-recovery-1.3206757).

Brym, Robert. 1983. "Israel in Lebanon." *Middle East Focus* 6: 14–19.

_________. 2007. "Six Lessons of Suicide Bombers." *Contexts* 8: 37–43.

_________. 2008. "Religion, Politics and Suicide Bombing: An Interpretative Essay." *Canadian Journal of Sociology* 33: 89–108.

_________. 2014. *2011 Census Update: A Critical Interpretation*. Toronto: Nelson.

Brym, Robert, and Robert Andersen. 2016. "Democracy, Women's Rights, and Public Opinion in Tunisia." *International Sociology* 31: 253–67

Brym, Robert, and Bader Araj. 2006. "Suicide Bombing as Strategy and Interaction: The Case of the Second *Intifada*." *Social Forces* 84: 1965–82.

_________ and Bader Araj. 2008. "Palestinian Suicide Bombing Revisited: A Critique of the Outbidding Thesis." *Political Science Quarterly* 123: 485–500.

_________ and Bader Araj. 2012a. "Are Suicide Bombers Suicidal?" *Studies in Conflict and Terrorism* (35, 6: 2012) 432–43.

_________ and Bader Araj. 2012b. "Suicidality and Suicide Bombers: A Rejoinder to Merari." *Studies in Conflict and Terrorism* 35: 733–39.

Brym, Robert, and Yael Maoz-Shai. 2009. "State Violence during the Second *Intifada*: Combining New Institutionalist and Rational Choice Approaches." *Studies in Conflict and Terrorism* 32: 611–26.

Bullard, Robert D. 2006. "Katrina and the Second Disaster: A Twenty-Point Plan to Destroy Black New Orleans." Retrieved May 18, 2006 (http://www.ejrc.cau.edu/Bullard20pointplan.html).

Burdeau, Cain. 2006. "Corps of Engineers Takes Responsibility for New Orleans Flooding." *Associated Press Newswires*, June 1.

Canadian Cancer Research Alliance. 2009. "Cancer Research Investment in Canada, 2007: The Canadian Cancer Research Alliance's Survey of Government and Voluntary Sector Investment in Cancer Research in 2007." Retrieved June 24, 2010 (http://www.ccra-acrc.ca/PDF%20Files/CCRA_EN_Main_2009.pdf).

Canadian Cancer Society. 2002. "Canadian Cancer Statistics." Retrieved June 22, 2010 (http://www.cancer.ca/~/media/CCS/Canada%20wide/Files%20List/English%20files%20heading/pdf%20not%20in%20publications%20section/2002%20Stats%20booklet%20%20En_72969285.ashx).

__________, Advisory Committee on Cancer Statistics. (2015). "Canadian Cancer Statistics 2015." Toronto: Canadian Cancer Society. Retrieved August 9, 2016 (http://www.cancer.ca/~/media/cancer.ca/CW/cancer%20information/cancer%20101/Canadian%20cancer%20statistics/Canadian-Cancer-Statistics-2015-EN.pdf?la=en).

"Canadian Jury Finds Afghan Family Guilty of 'Honour Killings.'" *The Guardian*, January 30. Retrieved April 23, 2013 (http://www.guardian.co.uk/world/2012/jan/30/honour-killings-jury-afghan-family).

Canadian Sociological Association and Department of Sociology, University of British Columbia. 2014. *Opportunities in Sociology*. Montreal.

Cancer Care Nova Scotia. 2006. "Understanding Cancer in Nova Scotia." Retrieved July 2, 2010 (http://www.cancercare.ns.ca/site-cc/media/cancercare/understandingcancer1.pdf).

Cancer Care Ontario. 2010. "Cancer Screening Completeness." Retrieved July 4, 2010 (http://csqi.cancercare.on.ca/cms/one.aspx?portalId=63405&pageId=67824).

Cancer Genome Atlas Network. 2008. "Comprehensive Genomic Characterization Defines Human Glioblastoma Genes and Core Pathways." *Nature*, October 23, pp. 1061–68.

Carreiro, Donna. 2016. "Indigenous children for sale: The money behind the Sixties Scoop." CBC News September 28. Retrieved December 30, 2016 (http://www.cbc.ca/news/canada/manitoba/sixties-scoop-americans-paid-thousands-indigenous-children-1.3781622)

CAREX Canada. 2010. "Surveillance of Environmental and Occupational Exposures for Cancer Prevention: Carcinogen Database." School of Environmental Health, University of British Columbia. Retrieved July 10, 2010 (http://www.carexcanada.ca/en/carcinogen_profiles_and_estimates).

"Carey Price." 2016. Spotrac. Retrieved December 30, 2016 (http://www.spotrac.com/nhl/montreal-canadiens/carey-price-1837).

"Celebrity 100 [The]." 2005. *Forbes.com*, June 15. Retrieved March 27, 2006 (http://www.forbes.com/celebrity100).

Centers for Disease Control and Prevention. 2016a. "Fatal Injury Reports, 1999–2014, for National, Regional, and States (RESTRICTED)." Retrieved August 8, 2016 (https://search.cdc.gov/search?query=Fatal+Injury+Reports%2C+1999%E2%80%922014&searchButton.x=0&searchButton.y=0&action=search&utf8=%E2%9C%93&affiliate=cdc-main).

_________. 2016b. "LCWK10. Deaths, percent of total deaths and rank order for 113 selected causes of death and Enterocolitis due to Clostridium difficile, by race and sex, United States, 2014." Retrieved August 8, 2016 (http://www.cdc.gov/nchs/data/dvs/lcwk10_2014.pdf).

Central Intelligence Agency. 2016. *The World Factbook*. Retrieved August 13, 2016 (https://www.cia.gov/library/publications/the-world-factbook).

Champion, Marc, and Lucette Lagnado. 2011. "A Tale of Two Alexandrias." *Wall Street Journal*, March 5. Retrieved April 22, 2013 (http://online.wsj.com/article/SB10001424052748703786804576138161911236484.html).

Charrad, M., and A. Zarrugh. 2014. "Equal or Complementary? Women in the New Tunisian constitution after the Arab Spring." *Journal of North African Studies* 19: 230–43.

Chen, Jing, Huixia Jiang, Bliss L. Tracy, and Jan M. Zielinski. 2008. "A Preliminary Radon Map for Canada According to Health Region." *Radiation Protection Dosimetry* 130: 92–94.

Chen, Lincoln C., Emdaqual Huq, and Stan D'Souza. 1981. "Sex Bias in the Family Allocation of Food and Health Care in Rural Bangladesh." *Population and Development Review* 7: 55–70

Chen, Yiqun. 2009. "Cancer Incidence in Fort Chipewyan, Alberta, 1995–2006." Alberta Cancer Board, Division of Population Health and Information Surveillance. Retrieved June 29, 2010 (http://www.albertahealthservices.ca/500.asp).

Chokshi, Niraj. 2015. "Map: The black homicide rate in (almost) every state." *Washington Post*, January 14. Retrieved August 8, 2016 (https://www.washingtonpost.com/blogs/govbeat/wp/2015/01/14/map-the-black-homicide-rate-in-almost-every-state).

Choo, Hae Yeon, and Myra Marx Ferree. 2010. "Practising Intersectionality in Sociological Research: A Critical Analysis of Inclusions, Interactions, and Institutions in the Study of Inequalities." *Sociological Theory* 28: 129–49.

Clark, Ronald W. 1971. *Einstein: The Life and Times*. New York: Avon.

Clement, W.H.P., 1897. *The History of the Dominion of Canada*. Toronto: William Briggs. The Copp, Clark Company. Retrieved December 30, 2016 (https://archive.org/stream/historyofdominio00clem#page/14/mode/2up).

Cohn, Marjorie. 2005. "The Two Americas." *Truthout*, September 3. Retrieved July 26, 2006 (http://www.truthout.org/docs_2005/090305Y.shtml).

Collyer, Dave. 2010. "Our World Has Changed." Presentation at the Calgary Chamber of Commerce, April 7. Retrieved June 30, 2010 (http://www.calgarychamber.com/resources/docs/David%20Collyer%20(CAPP)%20April%207%202010.pdf).

Combs, Sean "Puffy." 1999. *Forever*. New York: Bad Boy Entertainment (CD).

__________ and the Lox. 1997. "I Got the Power." Retrieved March 22, 2000 (http://www.ewsonline.com/badboy/lyrpow.html).

Comfort, Louise K. 2006. "Cities at Risk: Hurricane Katrina and the Drowning of New Orleans." *Urban Affairs Review* 41: 501–16.

Coser, Lewis. 1956. *The Functions of Social Conflict*. New York: Free Press.

da Vinci, Leonardo. 1956. "Of the Error Made by Those Who Practise without Science." P. 910 in *The Notebooks of Leonardo da Vinci*, translated and edited by Edward MacCurdy. New York: George Braziller.

Dale, Stephen Frederic. 1988. "Religious Suicide in Islamic Asia: Anticolonial Terrorism in India, Indonesia, and the Philippines." *Journal of Conflict Resolution* 32: 37–59.

Davies, Scott, and David Walters. 2011. "The Value of a Sociology Degree." Pp. 13–26 in *Society in Question*, 6th ed., edited by Robert J. Brym. Toronto: Nelson.

Davis, Joyce M. 2003. *Martyrs: Innocence, Vengeance and Despair in the Middle East*. New York: Palgrave Macmillan.

Davis, Mike. 1990. City of Quartz: Excavating the Future in Los Angeles. New York: Verso.

Dawsey, Darrell. 2006. "Proof's Death Fits Old Pattern." *Globe and Mail*, April 15. Retrieved April 15, 2006 (http://www.theglobeandmail.com/servlet/story/LAC.20060415.RAP15/TPStory/Entertainment).

"Delhi Gang-Rape Victim as Guilty as Her Rapists, Asaram Bapu Says." 2013. *Times of India*, January 7. Retrieved April 19, 2013 (http://articles.timesofindia.indiatimes.com/2013-01-07/india/36192073_1_asaram-bapu-sharma-and-akshay-thakur-chargesheet).

Denis, Jeffrey S. 2015. "Contact Theory in a Small-Town Settler–Colonial Context: The Reproduction of Laissez-Faire Racism in Indigenous–White Canadian Relations." *American Sociological Review* 80(1): 218–42.

__________. 2016. *Sociology of Indigenous People in Canada*, Robert Brym, ed. Toronto: Nelson.

DePratto, Brian. 2015. "Aboriginal Women Outperforming in Labour Markets." *TD Economics*, July 16. Retrieved December 30, 2016 (https://www.td.com/document/PDF/economics/special/AboriginalWomen.pdf).

Doberman, John. 1997. *Darwin's Athletes: How Sport Has Damaged Black America and Preserved the Myth of Race*. Boston: Houghton Mifflin.

Doll, Richard, and Richard Peto. 1981. *The Causes of Cancer: Qualitative Estimates of Avoidable Risks of Cancer in the United States Today*. New York: Oxford University Press.

Durkheim, Émile. 1951 [1897]. *Suicide: A Study in Sociology*, edited by George Simpson, translated by John A. Spaulding and George Simpson. New York: Free Press.

Dyson, Michael Eric. 2004. "The Culture of Hip-Hop." Pp. 61–68 in *That's the Joint! The Hip-Hop Studies Reader*, edited by Murray Forman and Mark Anthony Neal. New York: Routledge.

__________. 2006. *Come Hell or High Water: Hurricane Katrina and the Color of Disaster*. New York: Basic Civitas Books.

Edel, Abraham. 1965. "Social Science and Value: A Study in Interrelations." Pp. 218–38 in *The New Sociology: Essays in Social Science and Social Theory in Honor of C. Wright Mills*, edited by Irving Louis Horowitz. New York: Oxford University Press.

Egilman, David, and Samantha Howe. 2007. "Against Anti-Health Epidemiology: Corporate Obstruction of Public Health via Manipulation of Epidemiology." *International Journal of Occupational and Environmental Health* 13: 118–24.

Einstein, Albert. 1954. *Ideas and Opinions*, edited by Carl Seelig, translated by Sonja Bargmann. New York: Crown.

Eisner, Manuel, and Lana Ghuneim. 2013. "Honor Killing Attitudes amongst Adolescents in Amman, Jordan." *Aggressive Behavior* 39: 405–17.

Elran, Meir. 2006. *Khosen Leumi b'Yisrael: Hashpa'ot ha-Intifada ha-Shniya al ha-Khevra ha-Yisraelit.* [Hebrew: *National Resilience in Israel: The Influence of the Second Intifada on Israeli Society.*] Jaffee Center for Strategic Studies, University of Tel Aviv.

Emanuel, Kerry. 2005. "Increasing Destructiveness of Tropical Cyclones over the Past Thirty Years." *Nature* 436: 686–88.

Energy Information Administration, U.S. Department of Energy. 2005. "Hurricane Impacts on the U.S. Oil and Natural Gas Markets." Retrieved June 1, 2006 (http://tonto.eia.doe.gov/oog/special/eia1_katrina.html).

Energy Resources Conservation Board. 2009. "Directive 074." Retrieved June 29, 2010 (http://www.ercb.ca/docs/documents/directives/directive074.pdf).

Environics Institute. 2010. *Urban Aboriginal Peoples Study: Main Report.* Toronto. Retrieved December 30, 2016 (http://www.uaps.ca/wp-content/uploads/2010/04/UAPS-FULL-REPORT.pdf).

__________. 2016. *Canadian Public Opinion on Aboriginal Peoples.* Toronto. Retrieved December 30, 2016 (http://www.environicsinstitute.org/uploads/institute-projects/canadian%20public%20opinion%20on%20aboriginal%20peoples%202016%20-%20final%20report.pdf).

European Commission. 2010. "REACH—Registration, Evaluation, Authorisation and Restriction of Chemicals." Retrieved July 12, 2010 (http://ec.europa.eu/enterprise/sectors/chemicals/reach/index_en.htm).

Everett-Green, Robert. 1999. "Puff Daddy: The Martha Stewart of Hip Hop." *Globe and Mail,* September 4: C7.

"Facts about Lightning." 2006. *LEX18.com.* Retrieved April 29, 2006 (http://www.lex18.com/Global/story.asp?S=1367554&nav=menu203_3).

Fearon, E.R. 1997. "Human Cancer Syndrome: Clues to the Origin and Nature of Cancer." *Science* 278: 1043–50.

"50 Cent Slams Kanye's 'Bush Is Racist' Comment." 2005. *Contactmusic.com,* November 1. Retrieved May 23, 2006 (http://contactmusic.com/new/xmlfeed.nsf/mndwebpages/50%20cent%20slams%20kanyes%20bush%20is%20racist%20comment).

Forman, Murray. 2001. "It Ain't All about the Benjamins: Summit on Social Responsibility in the Hip-Hop Industry." *Journal of Popular Music Studies* 13: 117–23.

Francis, Daniel. 2011 [1992]. *The Imaginary Indian: The Image of the Indian in Canadian Culture*, 2nd ed. Vancouver: Arsenal Pulp Press.

Frank, Thomas, and Matt Weiland, eds. 1997. *Commodify Your Dissent: Salvos from the Baffler*. New York: W.W. Norton.

Frankl, Viktor E. 1959. *Man's Search for Meaning: An Introduction to Logotherapy*, translated by Ilse Lasch. Boston: Beacon Press.

Freedman, Jonathan L. 2002. *Media Violence and Its Effect on Aggression: Assessing the Scientific Evidence*. Toronto: University of Toronto Press.

Freedom House. 2006. "Freedom in the World 2006." Retrieved July 1, 2006 (http://www.freedomhouse.org/uploads/pdf/Charts 2006.pdf).

Friesen, Joe. 2013. "The future belongs to the young." *Globe and Mail*, January 19: A4

Galloway, Gloria. 2015. "70 per cent of murdered aboriginal women killed by indigenous men: RCMP." *Globe and Mail*, April 9. Retrieved December 30, 2016 (http://www.theglobeandmail.com/news/politics/70-per-cent-of-murdered-aboriginal-women-killed-by-indigenous-men-rcmp-confirms/article23868927).

Geddes, John. 2014. "Fort McKay First Nation chief made $644,441." *Maclean's*, September 25. Retrieved December 30, 2016 (http://www.macleans.ca/politics/ottawa/fort-mckay-first-nation-chief-made-644441).

George, Nelson. 1999. *Hip Hop America*. New York: Penguin.

Germanwatch. 2007. *The Climate Change Performance Index*. Bonn, Germany. Retrieved December 14, 2007 (http://www.germanwatch.org/ccpi/Klima/ccpi2008.pdf).

Ghosh, Bobby. 2011. "Rage, Rap and Revolution: Inside the Arab Youth Quake." *Time.com*, February 17. Retrieved December 30, 2016 (http://content.time.com/time/magazine/article/0,9171,2050022,00.html).

Gilbertson, Michael, and James Brophy. 2001. "Community Health Profile of Windsor, Ontario, Canada: Anatomy of a Great Lakes Area of Concern." *Environmental Health Perspectives* 109: 827–43.

Gill, Aisha K., Nazand Begikhani, and Gill Hague. 2012. "'Honour-Based Violence in Kurdish Communities." *Women's Studies International Forum* 35: 75–85.

Gilmore, Scott. 2015. "Canada's Race Problem? It's Even Worse than America's." *Maclean's*, January 22. Retrieved December 30, 2016 (http://www.macleans.ca/news/canada/out-of-sight-out-of-mind-2).

Glazier, R.H., M.I. Creatore, P. Gozdyra, F.I. Matheson, L.S. Steele, E. Boyle, and R. Moineddin. 2004. "Geographic Methods of Understanding and Responding to Disparities in Mammography Use in Toronto, Canada." *Journal of General Internal Medicine* 19: 952–61.

Goddard Institute for Space Studies. 2006. "Global Surface Air Temperature Anomaly (C) (Base: 1951–1980)." Retrieved April 24, 2006 (http://www.giss.nasa.gov/data/update/gistemp/graphs/Fig.A.txt).

__________. 2016. "GISS Surface Temperature Analysis (GISTEMP)". Retrieved August 8, 2016 (http://data.giss.nasa.gov/gistemp).

Government of Canada. 2002. "Study Released on Firearms in Canada." Retrieved December 29, 2005 (http://www.cfc-ccaf.gc.ca/media/news_releases/2002/survey-08202002_e.asp).

Gracyk, Theodore. 2001. *I Wanna Be Me: Rock Music and the Politics of Identity*. Philadelphia: Temple University Press.

Grant, Tavia. 2015. "Aboriginal women lead the way in Canada's labour markets." *Globe and Mail*, July 12. Retrieved December 30, 2016 (http://www.theglobeandmail.com/report-on-business/aboriginal-women-lead-the-way-in-canadas-labour-markets/article25475691).

Griner, Allison. 2013. "Aboriginal lawyers stride in footsteps of legal pioneer." Thunderbird.ca, March 25. Retrieved December 30, 2016 (http://thethunderbird.ca/2013/03/25/aboriginal-lawyers-stride-in-footsteps-of-legal-pioneer).

Groopman, Jerome. 2010. "The Plastic Panic." *New Yorker*, May 31. Retrieved July 8, 2010 (http://www.newyorker.com/reporting/2010/05/31/100531fa_fact_groopman).

Gross, Michael L. 2003. "Fighting by Other Means in the Mideast: A Critical Analysis of Israel's Assassination Policy."*Political Studies* 51: 350–68.

Gurr, Ted Robert. 1970. *Why Men Rebel*. Princeton, NJ: Princeton University Press.

"Haiti: Still Waiting for Recovery." 2013. *The Economist*, January 5. Retrieved April 28, 2013 (http://www.economist.com/news/americas/21569026-three-years-after-devastating-earthquake-republic-ngos-has-become-country).

Hajdu, David. 2005. "Guns and Poses." *New York Times*, March 11. Retrieved March 11, 2005 (www.nytimes.com).

Hall, Anthony J., and Gretchen Albers. 2015 [2011]. "Indigenous Peoples: Treaties." *Canadian Encyclopedia*. Retrieved December 30, 2016 (http://www.thecanadianencyclopedia.ca/en/article/aboriginal-treaties).

Hamilton, Maxwell J., Elaine de Valle, and Frances Robles. 2005. "Damage Extensive across Island." *Miami Herald*, July 10. Retrieved May 26, 2006 (http://www.latinamericanstudies.org/cuba/extensive.htm).

Hamlin, Cynthia, and Robert Brym. 2006. "The Return of the Native: A Cultural and Social-Psychological Critique of Durkheim's Suicide Based on the Guarani-Kaiowá of Southwestern Brazil." *Sociological Theory* 24: 42–57.

Hannah-Moffat, Kelly, and Pat O'Malley. 2007. *Gendered Risks*. Milton Park, UK: Routledge.

Harding, David J., Cybelle Fox, and Jal D. Mehta. 2002. "Studying Rare Events through Qualitative Case Studies: Lessons from a Study of Rampage School Shootings." *Sociological Methods and Research* 31: 174–217.

Harris, Gardiner. 2013. "India's New Focus on Rape Shows Only the Surface of Women's Perils." *New York Times*, January 12. Retrieved April 19, 2013 (http://www.nytimes.com/2013/01/13/world/asia/in-rapes-aftermath-india-debates-violence-against-women.html).

Henley, Jon. 2010. "Haiti: A Long Descent to Hell." *Guardian*, January 14. Retrieved April 28, 2013 (http://www.guardian.co.uk/world/2010/jan/14/haiti-history-earthquake-disaster).

Henrich, Joseph, Robert Boyd, and Peter J. Richerson. 2012. "The Puzzle of Monogamous Marriage." *Philosophical Transactions of the Royal Society: Biological Sciences* 367: 657–69. Retrieved April 23, 2013 (http://rstb.royalsocietypublishing.org/content/367/1589/657.full.pdf).

Henry, David A., Joel G. Ray, and Marcelo L. Urquia. 2012. "Sex Ratios among Canadian Liveborn Infants of Mothers from Different Countries." *Canadian Medical Association Journal* 184: E492–96.

Hill, A.B. 1965. "The Environment and Disease: Association or Causation?" *Proceedings of the Royal Society of Medicine* 68: 295–300.

Hirsch, Arnold R., and Joseph Logsdon. 1992. "Introduction to Part III." Pp. 189–200 in *Creole New Orleans: Race and Americanization*, edited by Arnold R. Hirsch and Joseph Logsdon. Baton Rouge, LA: Louisiana State University Press.

Holland, Greg J., and Peter J. Webster. 2007. "Heightened Tropical Cyclone Activity in the North Atlantic: Natural Variability or Climate Trend?" *Philosophical Transactions of the Royal Society A*, doi:10.1098/rsta.2007.2083. Retrieved August 25, 2007 (http://www.pubs.royalsoc.ac.uk/media/philtrans_a/Holland%20and%20Webster%201.pdf).

Holmes, Tamara E. 2005. "Blacks Underrepresented in Legal Field: ABA Report Shows Stark Contrasts in the Career Tracks of Lawyers." *Black Enterprise*, August. Retrieved April 30, 2005 (http://www.findarticles.com/p/articles/mi_m1365/is_1_36/ai_n15674277/pg_2).

Hoover, Robert N. 2000. "Cancer—Nature, Nurture, or Both." *New England Journal of Medicine* 343: 135–36.

House of Commons Canada. 2010. *Federal Poverty Reduction Plan: Working in Partnership towards Reducing Poverty in Canada: Report of the Standing Committee on Human Resources, Skills and Social Development and the Status of Persons with Disabilities*. Ottawa. Retrieved December 30, 2016 (http://www.parl.gc.ca/content/hoc/Committee/403/HUMA/Reports/RP4770921/humarp07/humarp07-e.pdf).

House, J. D. 2005. "Change from within the Corridors of Power: A Reflective Essay of a Sociologist in Government." *Canadian Journal of Sociology* 30: 281–314.

Huesmann, L. Rowell, Jessica Moise-Titus, Cheryl-Lynn Podolski, and Leonard D. Eron. 2003. "Longitudinal Relations between Children's Exposure to TV Violence and Their Aggressive and Violent Behavior in Young Adulthood: 1977–1992." *Developmental Psychology* 39: 201–21.

Hunnicutt, Gwen, and Kristy Humble Andrews. 2009. "Tragic Narratives in Popular Culture: Depictions of Homicide in Rap Music." *Sociological Forum* 24: 611–30.

Hunter, Shireen T. 1998. *The Future of Islam and the West: Clash of Civilizations or Peaceful Coexistence?* New York: Praeger.

Huntington, Samuel P. 1996. *The Clash of Civilizations and the Remaking of the World Order.* New York: Simon and Schuster.

Ignatieff, Michael. 2005. "The Broken Contract." *New York Times Magazine*, September 25, pp. 15–17.

Innis, Harold. 1977 [1930]. *The Fur Trade in Canada: An Introduction to Canadian Economic History*. Toronto: University of Toronto Press.

Intergovernmental Panel on Climate Change. 2007. *Climate Change 2007: The Physical Science Basis. Contribution of Working Group I to the Fourth Assessment Report of the Intergovernmental Panel on Climate Change*, edited by S. Solomon, D. Qin, M. Manning, Z. Chen, M. Marquis, K. B. Avery, M. Tignor, and H. L. Miller, Jr. Cambridge, UK: Cambridge University Press. Retrieved August 25, 2007 (http://ipcc-wg1.ucar.edu/wg1/Report/ AR4WG1_Pub_FAQs.pdf).

International Policy Institute for Counter-Terrorism. 2004. Retrieved November 1, 2004 (http://www.ict.org.il).

Israeli Ministry of Foreign Affairs. 2004. "Palestinian Violence and Terrorism since September 2000." Retrieved November 1, 2004 (http://www.mfa.gov.il/MFA/Terrorism-+Obstacle+to+Peace/Palestinian+terror+since+2000/Palestinian%20violence%20and%20terrorism%20since%20September).

James, William. 1976 [1902]. *The Varieties of Religious Experience: A Study in Human Nature.* New York: Collier Books.

Johnson, Jeffrey G., Patricia Cohen, Elizabeth M. Smailes, Stephanie Kasen, and Judith S. Brook. 2002. "Television Viewing and Aggressive Behavior during Adolescence and Adulthood." *Science* 295: 2468–71.

Johnson, S., L.T. McDonald, M. Corsten, and R. Rourke. 2010. "Socio-economic Status and Head and Neck Cancer Incidence in Canada: A Case-Control Study." *Oral Oncology* 46: 200–03.

Johnston, W. Robert. 2003. "Chronology of Terrorist Attacks in Israel, Part IV: 1993–2000." Retrieved October 25, 2004 (http://www.johnstonsarchive.net/terrorism/terrisrael-4.html).

Jones, Siân, Xiaosong Zhang, D. Williams Parsons, Jimmy Cheng-Ho Lin, Rebecca J. Leary, Philipp Angenendt, Parminder Mankoo, Hannah Carter, Hirohiko Kamiyama, Antonio Jimeno, Seung-Mo Hong, Baojin Fu, Ming-Tseh Lin, Eric S. Calhoun, Mihoko Kamiyama, Kimberly Walter, Tatiana Nikolskaya, Yuri Nikolsky,

James Hartigan, Douglas R. Smith, Manuel Hidalgo, Steven D. Leach, Alison P. Klein, Elizabeth M. Jaffee, Michael Goggins, Anirban Maitra, Christine Iacobuzio-Donahue, James R. Eshleman, Scott E. Kern, Ralph H. Hruban, Rachel Karchin, Nickolas Papadopoulos, Giovanni Parmigiani, Bert Vogelstein, Victor E. Velculescu, and Kenneth W. Kinzler. 2008. "Core Signaling Pathways in Human Pancreatic Cancers Revealed by Global Genomic Analyses." *Science*, September 26, pp. 1801–06.

Kandiyoti, Deniz. 1988. "Bargaining with Patriarchy." *Gender and Society* 2: 274–90.

Katz, Bruce, Matt Fellowes, and Mia Mabanta. 2006. *Katrina Index: Tracking Variables of Post-Katrina Reconstruction*. Washington, DC: The Brookings Institution. Retrieved April 24, 2006 (http://www.brookings.edu/metro/pubs/200604_KatrinaIndex.pdf).

Kaul, Rhythma. 2012. "Rape Victim Still Critical, Writes to Mother 'I Want to Live.'" *Hindustan Times*, December 20. Retrieved April 19, 2013 (http://www.hindustantimes.com/India-news/NewDelhi/Delhi-gang-rape-victim-writes-on-a-piece-of-paper-Mother-I-want-to-live/Article1-976798.aspx).

Keith, Margaret M., and James T. Brophy. 2004. "Participatory Mapping of Occupational Hazards and Disease among Asbestos-Exposed Workers from a Foundry and Insulation Complex in Canada." *International Journal of Occupational and Environmental Health* 10: 144–53.

Kelly, Erin N., David W. Schindler, Peter V. Hodson, Jeffrey W. Short, Roseanna Radamanovich, and Charlene C. Nielsen. 2010. "Oil Sands Development Contributes Elements Toxic at Low Concentrations to the Athabasca River and Its Tributaries." *Proceedings of the National Academy of Sciences* 107: 16178–83.

Kelly, Erin N., Jeffrey W. Short, David W. Schindler, Peter V. Hodson, Mingsheng Ma, Alvin K. Kwan, and Barbara L. Fortin. 2009. "Oil Sands Development Contributes Polycyclic Aromatic Compounds to the Athabasca River and Its Tributaries." *Proceedings of the National Academy of Sciences* 106: 22346–51.

Kevles, Daniel J. 1999. "Cancer: What Do They Know?" *New York Review of Books*, 46: 14–21.

Khan, Fawad. 2012. "In India, 22 Girls Get Kidnapped Every Day." *AAJ News*, August 11. Retrieved April 24, 2013 (http://www.aaj.tv/2012/08/in-india-22-girls-get-kidnapped-every-day).

King, G., and R. Bendel. 1995. "A Statistical Model Estimating the Number of African-American Physicians in the United States." *Journal of the National Medical Association* 87: 264–72.

King, Thomas. 2012. *The Inconvenient Indian: A Curious Account of Native People in North America.* Toronto: Doubleday Canada.

Knutson, Thomas R., and Robert E. Tuleya. 2004. "Impact of CO_2-Induced Warming on Simulated Hurricane Intensity and Precipitation: Sensitivity to the Choice of Climate Model and Convective Parameterization." *Journal of Climate* 17: 3477–95.

Kolbert, Elizabeth. 2006. *Field Notes from a Catastrophe.* London: Bloomsbury.

Korteweg, Anna, and Gökçe Yurdakul. 2010. "Religion, Culture and the Politicization of Honour-Related Violence: A Critical Analysis of Media and Policy Debates in Western Europe and North America." United Nations Research Institute for Social Development. Retrieved May 9, 2013 (http://www.unrisd.org/80256B3C005BCCF9/search/E61F80827BF3409FC1257744004DC465?OpenDocument=).

Krupa, Michelle. 2006. "Presumed Missing." *Times-Picayune*, March 6. Retrieved May 30, 2006 (http://www.nola.com/news/t-p/frontpage/index.ssf?/base/news-5/1141545589263750.xml).

Kurzweil, Ray. 1999. *The Age of Spiritual Machines: When Computers Exceed Human Intelligence.* New York: Viking Penguin.

Lapchick, Richard. 2016. "The Racial and Gender Report Card." Retrieved August 8, 2016 (http://www.tidesport.org/reports.html).

Laqueur, Walter. 2004. *No End to War: Terrorism in the Twenty-First Century.* New York: Continuum.

Law Society of British Columbia. 2015. "Quick Facts: About the Profession." Retrieved December 30, 2016 (https://www.lawsociety.bc.ca/page.cfm?cid=2189&t=About-the-Profession).

Lenin, Vladimir I. 1964. "Book Review: N. A. Rubakin, *Among Books*." Pp. 259–61 in Vol. 20, *Collected Works*, 45 vols. Moscow: Foreign Languages Publishing House.

Lenton, Rhonda L. 1989. "Homicide in Canada and the U.S.A." *Canadian Journal of Sociology* 14: 163–78.

Leon, D.A., D. Vågerö, and P.O. Olausson. 1992. "Social Class Differences in Infant Mortality in Sweden: Comparison with England and Wales." *British Medical Journal* 305: 687–91.

Lestschinsky, Jacob. 1948. *Crisis, Catastrophe and Survival: A Jewish Balance Sheet, 1914-1948*. New York: Institute of Jewish Affairs of the World Jewish Congress.

Li, Shuzhuo, Yan Wei, Quanbao Jiang, and Marcus W. Feldman. 2007. "Imbalanced Sex Ratio at Birth and Female Child Survival in China: Issues and Prospects." Pp. 25–47 in *Watering the Neighbour's Garden: The Growing Demographic Female Deficit in Asia*, edited by Isabelle Attané and Christophe Z. Guilmoto. Paris: Committee for International Cooperation in National Research in Demography. Retrieved April 24, 2013 (http://www.cicred.org/Eng/Publications/pdf/BOOK_ singapore.pdf).

Library of Congress. 2010. "Country Study: Maldives." Retrieved April 22, 2013 (http://lcweb2.loc.gov/frd/cs/mvtoc.html).

Lichtenstein, Paul, Niels V. Holm, Pia K. Verkasalo, Anastasia Iliadou, Jaakko Kaprio, Markku Koskenvuo, Eero Pukkala, Axel Skytthe, and Kari Hemminki. 2000. "Environment and Heritable Factors in the Causation of Cancer—Analyses of Cohorts of Twins from Sweden, Denmark, and Finland." *New England Journal of Medicine* 343: 78–85.

Liggio, John, Shao-Meng Li, Katherine Hayden, Youssef M. Taha, Craig Stroud, Andrea Darlington, Brian D. Drollette, Mark Gordon, Patrick Lee, Peter Liu, Amy Leithead, Samar G. Moussa, Danny Wang, Jason O'Brien, Richard L. Mittermeier, Jeffrey R. Brook, Gang Lu, Ralf M. Staebler, Yuemei Han, Travis W. Tokarek, Hans D. Osthoff, Paul A. Makar, Junhua Zhang, Desiree L. Plata, and Drew R. Gentner. 2016. "Oil Sands Operations as a Large Source of Secondary Organic Aerosols." *Nature* 534: 91–4. Retrieved August 10, 2016 (http://www.nature.com/nature/journal/v534/n7605/full/nature17646.html).

Link, Bruce G., and Jo Phelan. 1995. "Social Conditions as Fundamental Causes of Disease." *Journal of Health and Social Behavior* 35, Extra Issue: 80–94.

Logan, John R. 2006. "The Impact of Katrina: Race and Class in Storm-Damaged Neighborhoods." Department of Sociology, Brown University. Retrieved April 24, 2006 (http://www.s4.brown.edu/Katrina/report.pdf).

Logsdon, Joseph, and Caryn Cossé Bell. 1992. "The Americanization of Black New Orleans." Pp. 201–61 in *Creole New Orleans: Race and Americanization*, edited by Arnold R. Hirsch and Joseph Logsdon. Baton Rouge, LA: Louisiana State University Press.

Lukes, Steven. 1973. *Émile Durkheim, His Life and Work: A Historical and Critical Study*. London: Penguin.

MacCallum, Gerald C., Jr. 1967. "Negative and Positive Freedom." *The Philosophical Review* 76: 312–34.

Mahoney, Jill, 2013. "Canadians' attitudes hardening on aboriginal issues: new poll." *Globe and Mail* January 16, 2013. Retrieved December 30, 2016 (http://www.theglobeandmail.com/news/national/canadians-attitudes-hardening-on-aboriginal-issues-new-poll/article7408516).

Mahoney, Jill, and Alan Freeman. 2005. "Rebuilt City Likely to Be a Lot Smaller—and Whiter." *Globe and Mail*, September 17: A22. Toronto.

Mahony, Tina Hotton. 2010. "Police-Reported Dating Violence in Canada, 2008." *Juristat* 30. Retrieved April 24, 2013 (http://www.statcan.gc.ca/pub/85-002-x/2010002/article/11242-eng.pdf).

Mandel, N. 1965. "Turks, Arabs and Jewish Immigration into Palestine, 1882–1914." *St. Antony's Papers* 17: 77–108.

Mao, Yang, Jinfu Hu, Anne-Marie Ugnat, Robert Semenciw, Shirley Fincham, and the Canadian Cancer Registries Epidemiology Research Group. 2001. "Socioeconomic Status and Lung Cancer Risk in Canada." *International Journal of Epidemiology* 30: 809–17.

Margalit, Avishai. 2003. "The Suicide Bombers." *New York Review of Books*, 50. Retrieved September 1, 2004 (http://www.nybooks.com/articles/15979).

Marshall, Tabitha. 2014 [2013]. "Oka Crisis." *Canadian Encyclopedia*. Retrieved December 30, 2016 (http://www.thecanadianencyclopedia.ca/en/article/oka-crisis).

Marshall, T.H. 1965. "Citizenship and Social Class." Pp. 71–134 in *Class, Citizenship, and Social Development: Essays by T H. Marshall*, edited by T.H. Marshall. Garden City, NY: Anchor.

Martin, Susan Taylor. 2005. "Can We Learn from Cuba's Lesson?" *St. Petersburg Times*, September 9. Retrieved May 26, 2006 (http://www.sptimes.com/2005/09/09/Worldandnation/Can_we_learn_from_Cub.shtml).

Mattern, Mark. 1998. *Acting in Concert: Music, Community, and Political Action*. New Brunswick, NJ: Rutgers University Press.

McWhorter, John. 2005. *Winning the Race: Beyond the Crisis in Black America*. New York: Gotham Books.

Mead, George Herbert. 1934. *Mind, Self and Society*. Chicago: University of Chicago Press.

Miall, Andrew D. 2013. "The Environmental Hydrogeology of the Oil Sands, Lower Athabasca Area, Alberta." *Geoscience Canada* 40: 215–33

Miller, Barbara. 2010. "Daughter Neglect, Women's Work, and Marriage: Pakistan and Bangladesh Compared." *Medical Anthropology* 8: 109–26

Miller, James R., and Zach Parrott. 2015 [2006]. "Indigenous–British Relations Pre-Confederation." *Canadian Encyclopedia*. Retrieved December 30, 2016 (http://www.thecanadianencyclopedia.com/en/article/aboriginal-european-relations).

Milloy, John. 1996. "Residential Schools." Pp. 309–94 in *Royal Commission on Aboriginal Peoples*. Ottawa: Government of Canada. Retrieved December 30, 2016 (https://qspace.library.queensu.ca/bitstream/1974/6874/5/RRCAP1_combined.pdf).

Mills, C. Wright. 1959. *The Sociological Imagination*. New York: Oxford University Press.

Moghissi, Haideh. 1999. *Feminism and Islamic Fundamentalism: The Limits of Postmodern Analysis*. London: Zed Books.

Moore, Barrington, Jr. 1967. *Social Origins of Dictatorship and Democracy: Lord and Peasant in the Making of the Modern World*. Boston: Beacon Press.

Mosby, Ian. 2013. "Administering Colonial Science: Nutrition Research and Human Biomedical Experimentation in Aboriginal Communities and Residential Schools, 1942–1952." *Histoire Sociale/Social History* 46(91): 145–72.

Moses, John. 2013. "Lives Lived: Russell Copeland Moses, CD, 80." *Globe and Mail*, July 11. Retrieved December 30, 2016 (http://www.theglobeandmail.com/life/facts-and-arguments/lives-lived-russell-copeland-moses-cd-80/article13125778).

Moses, Russ. 1965. Letter to L. Jampolsky. Retrieved December 30, 2016 (http://beta.images.theglobeandmail.com/static/TRC/RussMoses.pdf).

National Center for Education Statistics. 2004. "Table 303. Degrees Awarded by Degree-Granting Institutions, by Control, Level of Degree, and State or Jurisdiction: 2002–03." Retrieved July 2, 2006 (http://nces.ed.gov/programs/digest/d04/tables/dt04_303.asp).

National Center for Injury Prevention and Control. 2016. "Leading Causes of Death Reports, 1999–2015, for National, Regional, and

States (RESTRICTED)." Retrieved December 27, 2016. (https://webappa.cdc.gov/cgi-bin/broker.exe).

National Rifle Association. 2005. "Guns, Gun Ownership, & RTC at All-Time Highs, Less 'Gun Control,' and Violent Crime at 30-Year Low." Retrieved December 29, 2005 (http://www.nraila.org/Issues/FactSheets/Read.aspx?ID=126).

National Weather Service. 2005. "The Saffir–Simpson Hurricane Scale." Retrieved June 2, 2006 (http://www.nhc.noaa.gov/aboutsshs.shtml).

Nayak, Madhabika B., Christina A. Byrne, Mutsumi K. Martin, and Ann George Abraham. 2003. "Attitudes toward Violence against Women: A Cross-Nation Study." *Sex Roles* 49: 333–42.

Neal, Mark Anthony. 1999. *What the Music Said: Black Popular Music and Black Public Culture.* New York: Routledge.

New York Times. 2000–2006. East Coast Final Edition. New York.

Nicole (Jessi Mechler). August 15, 2014. *Boyhood.* IFC Productions, Detour Filmproduction.

Nikiforuk, Andrew. 2008. *Tar Sands: Dirty Oil and the Future of a Continent.* Vancouver: Greystone.

Nordheimer, Jon. 2002. "Nothing's Easy for New Orleans Flood Control." *New York Times*, April 30: F1. New York.

Obermeyer, Carla Makhlouf, Eva Deykin, and Joseph Potter. 1993. "Paediatric Care and Immunisation among Jordanian Children." *Journal of Biosocial Science* 25: 371–81

Obermeyer, Carla Makhlouf, and Rosario Carednas. 1997. "Son Preference and Differential Treatment in Morocco and Tunisia." *Studies in Family Planning.* 28: 235–44

Obomsawin, Alanis. 1986. *Cry from a Diary of a Metis Child.* Toronto: National Film Board. (film)

O'Donnell, Vivian, and Susan Wallace. 2011. "First Nations, Métis and Inuit Women." *Statistics Canada.* Retrieved May 31, 2013 (http://www.statcan.gc.ca/pub/89-503-x/2010001/article/11442-eng.htm).

Oliver, Anne Marie, and Paul Steinberg. 2005. *The Road to Martyrs' Square: A Journey into the World of the Suicide Bomber.* Oxford, UK: Oxford University Press.

"Once upon a Time in Egypt." 2011. *Foreign Policy*, April 25. Retrieved April 22, 2013 (http://www.foreignpolicy.com/articles/2011/04/25/once_upon_a_time_in_egypt#2).

Oreskes, Naomi. 2004. "The Scientific Consensus on Climate Change." *Science*, December 3, p. 1686.

Pape, Robert A. 2005. *Dying to Win: The Strategic Logic of Suicide Terrorism.* New York: Random House.

Parsons, D. Williams, Siân Jones, Xiaosong Zhang, Jimmy Cheng-Ho Lin, Rebecca J. Leary, Philipp Angenendt, Parminder Mankoo, Hannah Carter, I-Mei Siu, Gary L. Gallia, Alessandro Olivi, Roger McLendon, B. Ahmed Rasheed, Stephen Keir, Tatiana Nikolskaya, Yuri Nikolsky, Dana A. Busam, Hanna Tekleab, Luis A. Diaz, Jr., James Hartigan, Doug R. Smith, Robert L. Strausberg, Suely Kazue Nagahashi Marie, Sueli Mieko Oba Shinjo, Hai Yan, Gregory J. Riggins, Darell D. Bigner, Rachel Karchin, Nick Papadopoulos, Giovanni Parmigiani, Bert Vogelstein, Victor E. Velculescu, and Kenneth W. Kinzler. "An Integrated Genomic Analysis of Human Glioblastoma Multiforme." 2008. *Science*, September 26, pp. 1807–12.

Pastore, Ralph. 1997. "The Beothuk." *Heritage Newfoundland and Labrador*. Retrieved December 30, 2016 (http://www.heritage.nf.ca/articles/aboriginal/beothuk.php).

Patel, Sujay, and Amin Muhammad Gadit. 2008. "Karo-Kari: A Form of Honour Killing in Pakistan." *Transcultural Psychiatry* 45: 683–94.

Pembina Institute. 2007. "Albertans' Perceptions of Oil Sands Development: Poll Part 1: Pace and Scale of Oil Sands Development." Retrieved June 28, 2013 (http://pubs.pembina.org/reports/Poll_Env_mediaBG_Final.pdf).

Percy, Kevin E. 2013. "Ambient Air Quality and Linkage to Ecosystems in the Athabasca Oil Sands, Alberta." *Geoscience Canada* 40: 182–201

Perina, Kaja. 2002. "Suicide Terrorism: Seeking Motives beyond Mental Illness." *Psychology Today* 35: 15.

Perlman, David. 2010. "Port-au-Prince Buildings Poorly Reinforced." *San Francisco Chronicle*, January 27. Retrieved April 28, 2013 (http://www.sfgate.com/news/article/Port-au-Prince-buildings-poorly-reinforced-3274772.php).

Perreault, Samuel, and Shannon Brennan. 2010. "Criminal Victimization in Canada, 2009." Statistics Canada. Retrieved April 25, 2013 (http://www.statcan.gc.ca/pub/85-002-x/2010002/article/11340-eng.htm).

Pettigrew, Thomas F., and Linda R. Tropp. 2006. "A Meta-Analytic Test of Intergroup Contact Theory." *Journal of Personality and Social Psychology* 90(5): 751–83.

Pew Forum on Religion and Public Life. 2010. "The Future of the Global Muslim Population." Retrieved April 21, 2013 (http://features.pewforum.org/muslim-population).

Phelan, Jo C., Bruce G. Link, Ana Diez-Roux, Ichiro Kawachi, and Bruce Levin. 2004. "'Fundamental Causes' of Social Inequalities in Mortality: A Test of the Theory." *Journal of Health and Social Behavior* 45: 265–85.

Physicians for a Smoke-Free Canada. 2009. "Smoking in Canada." Retrieved June 21, 2010 (http://www.smoke-free.ca/factsheets/pdf/prevalence.pdf).

_________. 2012. "Factsheets." Retrieved August 9, 2016 (http://www.smoke-free.ca/factsheets/default.htm).

Piven, Frances Fox, and Richard A. Cloward. 1977. *Poor People's Movements: Why They Succeed, How They Fail.* New York: Vintage.

_________. 1993. *Regulating the Poor: The Functions of Public Welfare*, Updated edition. New York: Vintage.

Polanyi, Karl. 1957. *The Great Transformation: The Political and Economic Origins of Our Time.* Boston: Beacon Press.

Pruitt, Sandi L., Matthew J. Shim, Patricia Dolan Mullen, Sally W. Vernon, and Benjamin C. Amick III. 2009. "Association of Area Socioeconomic Status and Breast, Cervical, and Colorectal Cancer Screening: A Systematic Review." *Cancer Epidemiology, Biomarkers & Prevention* 18: 2579–99.

Ramos, Howard. 2006. "What Causes Canadian Aboriginal Protest? Examining Resources, Opportunities and Identity, 1951–2000." *Canadian Journal of Sociology* 31(2): 211–34.

_________. 2008. "Opportunity for Whom? Political Opportunity and Critical Events in Canadian Aboriginal Mobilization, 1951–2000." *Social Forces* 87(2): 795–823.

Remennick, Larissa I. 1998. "The Cancer Problem in the Context of Modernity: Sociology, Demography, Politics." *Current Sociology* 46: 1–150.

Reuter, Christoph. 2004. *My Life Is a Weapon: A Modern History of Suicide Bombing*, translated by Helena Ragg-Kirkby. Princeton, NJ: Princeton University Press.

Reuters News Agency. 2005. "Hurricane Dennis Killed 16 in Cuba—Castro." Retrieved May 26, 2006 (http://www.planetark.org/dailynewsstory.cfm/newsid/31634/newsDate/13-Jul-2005/story.htm).

Ricolfi, Luca. 2005. "Palestinians, 1981–2003." Pp. 77–129 in *Making Sense of Suicide Missions*, edited by Diego Gambetta. Oxford: Oxford University Press.

Risks of Tobacco Smoking and Toxic Constituents of Smoke: Results from the 2002 International Tobacco Control (ITC) Four Country Survey." *Tobacco Control* 15. Retrieved July 4, 2010 (http://tobaccocontrol.bmj.com/content/15/suppl_3/iii65.full).

Robertoux, Pierre L., and Michèle Carlier. 2007. "From DNA to Mind." *European Molecular Biology Reports* 8: S7–S11.

Romero, Simon. 2013. "Public Rapes Outrage Brazil, Testing Ideas of Image and Class." *New York Times*, May 24. Retrieved May 24, 2013 (http://www.nytimes.com/2013/05/25/world/americas/rapes-in-brazil-spur-class-and-gender-debate.html?pagewanted=all&_r=0).

Samson, Colin, James Wilson, and Jonathan Mazower. 1999. *Canada's Tibet: The Killing of the Innu*. London UK: Survival International. Retrieved December 30, 2016 (http://assets.survivalinternational.org/static/files/books/InnuReport.pdf).

Samuels, David. 2004. "The Rap on Rap: The 'Black Music' That Isn't Either." Pp. 147–53 in *That's the Joint! The Hip-Hop Studies Reader*, edited by Murray Forman and Mark Anthony Neal. New York: Routledge.

Sarria, Nidya. 2009. "Femicides of Juárez: Violence against Women in Mexico." *Council on Hemispheric Affairs.* Retrieved May 30, 2013 (http://www.coha.org/femicides-of-juarez-violence-against-women-in-mexico).

Saul, Stephanie. 2010. "Earliest Steps to Find Breast Cancer Are Prone to Error." *New York Times*, July 20. Retrieved July 20, 2010 (http://www.nytimes.com).

Schiermeier, Quirin. 2005a. "Hurricane Link to Climate Change Is Hazy." *Nature* 437: 461.

_________. 2005b. "Trouble Brews over Contested Trend in Hurricanes." *Nature* 435: 1008–09.

Schindler, David W. 2013. "Water Quality Issues in the Oil Sands Region of the Lower Athabasca River, Alberta." *Geoscience Canada* 40: 202–14

Schwartz, John. 2007. "One Billion Dollars Later, New Orleans Is Still at Risk." *New York Times*, August 17. Retrieved August 17, 2007 (http://www.nytimes.com).

Sexton, Richard, and Randolph Delehanty. 1993. *New Orleans: Elegance and Decadence*. San Francisco: Chronicle Books.

Siahpush, M., A. McNeill, D. Hammond, and G.T. Fong. 2006. "Socioeconomic and country variations in knowledge of health risks of tobacco smoking and toxic constituents of smoke: Results from the 2002 International Tobacco Control (ITC) Four Country Survey." U.S. National Library of Medicine. BMJ Publishing Group Ltd.

Silverman, Adam L. 2002. "Just War, Jihad, and Terrorism: A Comparison of Western and Islamic Norms for the Use of Political Violence." *Journal of Church and State* 44: 73–92.

Simieritsch, Terra, Joe Obad, and Simon Dyer. 2009. Tailings Plan Review—An Assessment of Oil Sands Company Submissions for Compliance with ERCB Directive 074: Tailings Performance Criteria and Requirements for Oil Sands Mining Schemes. Drayton Valley and Canmore, AB: Pembina Institute and Water Matters. Retrieved June 29, 2010 (http://pubs.pembina.org/reports/tailings-plan-review-report.pdf).

Simmel, Georg. 1950. *The Sociology of Georg Simmel*, translated and edited by Kurt H. Wolff. New York: Free Press.

Sinhai, Kounteya. 2012. "Govt Mulls Ban on Portable Ultrasound Machines." *Times of India*, January 7. Retrieved April 20, 2013 (http://articles.timesofindia.indiatimes.com/2012-01-07/india/30601426_1_ultrasound-machines-csb-mobile-clinic).

Spalter-Roth, Roberta, William Erskine, Sylvia Polsiak, and Jamie Panzarella. 2005. *A National Survey of Seniors Majoring in Sociology*. Washington, DC: American Sociological Association. Retrieved December 12, 2005 (http://www.asanet.org/galleries/default-file/B&B_first_report_final.pdf).

Springhall, John. 1998. *Youth, Popular Culture and Moral Panics: Penny Gaffs to Gangsta-Rap, 1830–1996.* New York: Routledge.

Sprinzak, Ehud. 2000. "Rational Fanatics." *Foreign Policy* 120: 66–73.

Statistics Canada. 2011. "Family Violence in Canada: A Statistical Profile." Retrieved April 25, 2013 (http://www.statcan.gc.ca/pub/85-224-x/85-224-x2010000-eng.pdf).

_________. 2015a. "The educational attainment of Aboriginal peoples in Canada." Retrieved December 30, 2016 (https://www12.statcan.gc.ca/nhs-enm/2011/as-sa/99-012-x/99-012-x2011003_3-eng.cfm).

_________. 2015b. "Aboriginal Peoples in Canada: First Nations People, Métis and Inuit." Retrieved December 30, 2016 (https://www12.statcan.gc.ca/nhs-enm/2011/as-sa/99-011-x/99-011-x2011001-eng.cfm#a2).

_________. 2016. "Aboriginal peoples of Canada." Retrieved December 30, 2016 (http://www12.statcan.ca/english/census01/Products/Analytic/companion/abor/canada.cfm).

Statistics Centre—Abu Dhabi. 2016. "Population and Demography." Retrieved August 13, 2016 (https://www.scad.ae/en/Pages/ThemesReleases.aspx?ThemeID=4).

"Stelmach Prepares to Take Charge as Premier." 2006. Retrieved June 30, 2010 (http://www.cbc.ca/canada/edmonton/story/2006/12/04/stelmach-monday.html).

Stephens, W. Richard, Jr. 1998. *Careers in Sociology*. New York: Allyn & Bacon. Retrieved July 3, 2006 (http://www.abacon.com/socsite/careers.html).

Stern, Jessica. 2003. *Terror in the Name of God: Why Religious Militants Kill.* New York: Ecco/HarperCollins.

Stockett, Kathryn. 2009. *The Help*. New York: Berkley.

Strong, Nolan. 2006. "Lil Kim's Reality Show Scores Highest Debut in BET History." *Allhiphop.com*, March 14. Retrieved April 23, 2006 (http://www.allhiphop.com/hiphopnews/?ID=5460).

Sullivan, Mercer L. 2002. "Exploring Layers: Extended Case Method as a Tool for Multilevel Analysis of School Violence." *Sociological Methods and Research* 31: 255–85.

"Survey of Hurricane Katrina Evacuees." 2005. *Washington Post* and the Henry J. Kaiser Family Foundation and the Harvard School of Public Health. Retrieved April 24, 2006 (http://www.washingtonpost.com/wp-srv/politics/polls/katrina_poll091605.pdf).

Svenson, Larry. 2010. Personal correspondence, June 29. [Svenson is director, Epidemiology and Surveillance, Surveillance and Assessment Branch, Community and Population Health, Alberta Health and Wellness.]

Swidler, Ann. 1986. "Culture in Action: Symbols and Strategies." *American Sociological Review* 51: 273–86.

Taarnby, Michael. 2003. *Profiling Islamic Suicide Terrorists: A Research Report for the Danish Ministry of Justice*. Aarhus, Denmark: Centre for Cultural Research, University of Aarhus. Retrieved September 1, 2004 (http://www.jm.dk/image.asp?page=image&objno=71157).

Tanner, Julian, Mark Asbridge, and Scot Wortley. 2009. "Listening to Rap: Cultures of Crime, Cultures of Resistance." *Social Forces* 88: 693–722.

Taylor, Charles L. 1985. "What's Wrong with Negative Liberty?" Pp. 211–29 in *Philosophy and Human Sciences: Philosophical Papers 2*. Cambridge: Cambridge University Press.

Tidwell, Mark. 2004. *Bayou Farewell: The Rich Life and Tragic Death of Louisiana's Cajun Coast*. New York: Vintage.

_________. 2005. "It's Time to Abandon New Orleans." *Winnipeg Free Press*, December 9: A15. Winnipeg.

Timoney, Kevin P. 2007. "A Study of Water and Sediment Quality as Related to Public Health Issues, Fort Chipewyan, Alberta." Sherwood Park, AB. Retrieved June 28, 2010 (http://www.connectingthedrops.ca/docs/fc-final-report-revised-dec2007.pdf).

Timoney, Kevin P., and Peter Lee. 2009. "Does the Alberta Tar Sands Industry Pollute? The Scientific Evidence." *The Open Conservation Biology Journal* 3: 65–81. Retrieved June 22, 2010 (http://www.bentham.org/open/toconsbj/openaccess2.htm).

"Top 100 Richest Rappers." 2016. Retrieved August 8, 2016 (http://www.therichest.com/top-lists/top-100-richest-rappers).

Toxic Trespass. 2006. Toronto: National Film Board of Canada.

Treasury Board of Canada Secretariat. 2016. *Federal Contaminated Sites Inventory*. Retrieved August 19, 2016 (http://www.tbs-sct.gc.ca/fcsi-rscf/home-accueil-eng.aspx).

Truth and Reconciliation Commission. 2012. *They Came for the Children: Canada, Aboriginal Peoples, and Residential School*. Winnipeg: Truth and Reconciliation Commission of Canada. Retrieved December 30, 2016 (http://www.myrobust.com/websites/trcinstitution/File/2039_T&R_eng_web%5B1%5D.pdf).

"UAE has high male to female population ratio." *UAE News General*, August 19, 2002. Retrieved August 14, 2016 (http://gulfnews.com/news/uae/general/uae-has-high-male-to-female-population-ratio-1.396683).

UNESCO. 2010. "Education under Attack 2010—Afghanistan." Retrieved May 4, 2013 (http://www.refworld.org/cgi-bin/texis/vtx/rwmain?page=publisher&docid=4b7aa9e6c&skip=0&publisher=UNESCO&querysi=Education%20under%20attack%202010%20-%20Afghanistan&searchin=title&sort=date).

United Nations. 1946. "Resolution 96 (I): The Crime of Genocide." *Resolutions Adopted by the General Assembly during its First Session*. Retrieved December 30, 2016 (http://www.un.org/documents/ga/res/1/ares1.htm).

__________. 2004a. "Cuba: A Model in Hurricane Risk Management." Retrieved May 26, 2006 (http://www.un.org/News/Press/docs/2004/iha943.doc.htm).

__________. 2004b. "Reducing Disaster Risk: A Challenge for Development. New York." Retrieved May 18, 2006 (http://www.undp.org/bcpr/disred/documents/publications/rdr/english/rdr_english.pdf).

__________. 2005. "Human Development Report 2005." Retrieved July 1, 2006 (http://hdr.undp.org/reports/global/2005).

__________. 2010. "Table 4: Gender Inequality Index." Retrieved April 21, 2013 (http://hdr.undp.org/en/media/HDR_2010_EN_Table4_reprint.pdf).

__________. 2013a. *Human Development Report 2013.* Retrieved April 23, 2013 (http://hdr.undp.org/hdr4press/press/report/hdr/english/HDR2013_EN_Complete.pdf).

__________. 2013b. "UN Women in Maldives." Retrieved April 21, 2013 (http://www.unwomensouthasia.org/un-women-in-south-asia-2/un-women-in-maldives).

U.S. Census Bureau. 2002. "Table 1. United States—Race and Hispanic Origin: 1790 to 1990." Retrieved April 29, 2006 (http://www.census.gov/population/documentation/twps0056/tab01.xls).

__________. 2010. "Table 4. Annual Estimates of the Black or African American Alone or in Combination Resident Population by Sex and Age for the United States: April 1, 2000 to July 1, 2009." Retrieved July 20, 2010 (http://www.census.gov/popest/national/asrh/NC-EST2009/NC-EST2009-04-BAC.xls).

__________. 2005. "Hurricane Katrina Disaster Areas." Retrieved May 30, 2006 (http://ftp2.census.gov/geo/maps/special/HurKat/Katrina_Reference_v2.pdf).

__________. 2006. "Entire Data Set." Retrieved April 15, 2006. (http://www.census.gov/programs-surveys/popest.html).

__________. 2016a. "The Black Alone Population in the United States: 2013." Retrieved August 8, 2016 (http://www.census.gov/population/race/data/ppl-ba13.html).

__________. 2016b. "Annual Estimates of the Resident Population by Sex, Race, and Hispanic Origin for the United States, States, and Counties: April 1, 2010 to July 1, 2015." Retrieved August 8, 2016 (https://www.census.gov/popest/data/datasets.html).

__________. 2016c. "New Orleans city, Louisiana." Retrieved August 8, 2016 (http://www.census.gov/quickfacts/table/PST045215/2255000).

U.S. Department of Health, Education, and Welfare. 1964. *Smoking and Health: Report of the Advisory Committee to the Surgeon General of the Public Health Service*. Washington, DC. Retrieved June 21, 2010 (http://profiles.nlm.nih.gov/NN/B/B/M/Q/_/nnbbmq.pdf).

U.S. Department of Health and Human Services. 2009. *Report on Carcinogens*, 11th edition. Public Health Service, National Toxicology Program. Retrieved June 25, 2010 (http://ntp.niehs.nih.gov/index.cfm?objectid=32BA9724-F1F6-975E-7FCE50709CB4C932).

U.S. Department of Labor. Bureau of Labor Statistics. 2015. "May 2015 National Occupational Employment and Wage Estimates." Retrieved December 24, 2016 (https://www.bls.gov/oes/current/oes_nat.htm#00-0000).

U.S. Energy Information Administration. 2009. "World Proved Reserves of Oil and Natural Gas, Most Recent Estimates." Retrieved June 27, 2010 (http://www.eia.doe.gov/emeu/international/reserves.html).

U.S. House of Representatives, Select Bipartisan Committee to Investigate the Preparation for and Response to Hurricane Katrina. 2005. *A Failure of Initiative: The Final Report of the Select Bipartisan Committee to Investigate the Preparation for and Response to Hurricane Katrina.* Retrieved April 24, 2006 (http://katrina.house.gov).

Vanderklippe, Nathan. 2012. "In oil sands, a native millionaire sees 'economic force' for first nations." *Globe and Mail*, August 13. Retrieved December 30, 2016 (http://www.theglobeandmail.com/report-on-business/industry-news/energy-and-resources/in-oil-sands-a-native-millionaire-sees-economic-force-for-first-nations/article4479795).

Victor, Barbara. 2003. *Army of Roses: Inside the World of Palestinian Women Suicide Bombers.* New York: Rodale Press.

Voyageur, Cora J., and Brian Calliou. 2000–01. "Various Shades of Red: Diversity within Canada's Indigenous Community." *London Journal of Canadian Studies* 16: 103–18.

Wade, Nicholas. 2010. "A Decade Later, Genetic Map Yields Few New Cures." *New York Times*, June 12. Retrieved June 12, 2010 (http://www.nytimes.com).

Wallace, Ron R. 2013. "History and Governance Models as a Blueprint for Future Federal–Provincial Co-operation on Environmental Monitoring in the Alberta Oil Sands." *Geoscience Canada* 40: 167–73.

Wallerstein, Immanuel. 1974–1989. *The Modern World-System*, 3 vols. New York: Academic Press.

Walsh, Declan. 2012. "Taliban Gun Down Girl Who Spoke Up for Rights." *New York Times*, October 9. Retrieved May 9, 2013 (http://www.nytimes.com/2012/10/10/world/asia/teen-school-activist-malala-yousafzai-survives-hit-by-pakistani-taliban.html?pagewanted=all&_r=0).

"Washing Away: Special Report from *Times-Picayune*," June 23–27, 2002. Retrieved June 1, 2006 (http://www.nola.com/hurricane/?/washingaway).

Weber, Max. 1946. "Class, Status, Party." Pp. 180–95 in *From Max Weber: Essays in Sociology*, translated and edited by Hans Gerth and C. Wright Mills. New York: Oxford University Press.

_________. 1947. *The Theory of Social and Economic Organization*, edited by T. Parsons, translated by A. M. Henderson and T. Parsons. New York: Free Press.

_________. 1964. "'Objectivity' in Social Science and Social Policy." Pp. 49–112 in *The Methodology of the Social Sciences*, translated and edited by Edward A. Shils and Henry A. Finch. New York: Free Press of Glencoe.

Webster, P. J., G. J. Holland, J. A. Curry, and H.-R. Chang. 2005. "Changes in Tropical Cyclone Number, Duration, and Intensity in a Warming Environment." *Science* 309: 1844–46.

Weitzer, Ronald, and Charis E. Kubrin. 2009. "Misogyny in Rap Music: A Content Analysis of Prevalence and Meanings." *Men and Masculinities* 12: 3–29.

Welch, Michael Dylan. 2016. "The Tiny Room: The Jottings of E.E. Cummings." Retrieved December 30, 2016 (http://www.graceguts.com/essays/the-tiny-room-the-jottings-of-e-e-cummings).

Weldon, S. Laurel. 2002. *Protest, Policy, and the Problem of Violence against Women: A Cross-National Comparison.* Pittsburgh: University of Pittsburgh Press.

Wells, George. 2010. "Standardization (of Rates)." *Encyclopedia of Public Health.* Retrieved June 22, 2010 (http://www.enotes.com/public-health-encyclopedia/standardization-rates).

"Why So Much Medical Research Is Rot." 2007. *Cancer World,* May/June. Retrieved July 7, 2010 (http://www.cancerworld.org/pdf/5370_CW18_62_63_focus.pdf).

Wikipedia. 2016a. "List of murdered hip hop musicians." Retrieved August 8, 2016. (https://en.wikipedia.org/wiki/List_of_deceased_hip_hop_artists).

__________. 2016b. "List of black NHL players." Retrieved August 8, 2016 (https://en.wikipedia.org/wiki/List_of_black_NHL_players).

Wilson, William Julius. 1987. *The Truly Disadvantaged: The Inner City, the Underclass, and Public Policy.* Chicago: University of Chicago Press.

Wiseman, Richelle. 2010. "The Honour Killing Debate in Canada." Paper presented at "Gender, Culture and Religion," a symposium sponsored by the Sheldon Chumir Foundation for Ethics in Leadership (Calgary). Retrieved April 23, 2013 (http://www.chumirethicsfoundation.ca/files/pdf/HonourKillingsPaper_rw_20100827_web.pdf).

WomanStats Project. 2013. "Women's Physical Security." Retrieved April 20, 2013 (http://womanstats.org/substatics/Women%27s%20Physical%20Security_2009tif_wmlogo3.png).

"Women and Islam." 2013. *Oxford Islamic Studies Online.* Retrieved April 22, 2013 (http://www.oxfordislamicstudies.com/article/opr/t125/e2510?_hi=6&_pos=2).

World Bank. 2013. "GDP per Capita (Current US$)." Retrieved April 21, 2013 (http://data.worldbank.org/indicator/NY.GDP.PCAP.CD).

World Health Organization. 2002. "The Tobacco Atlas." Geneva. Retrieved July 24, 2010 (http://whqlibdoc.who.int/publications/2002/9241562099.pdf).

Wotherspoon, Terry, and John Hansen. 2013. "The 'Idle No More' Movement: Paradoxes of First Nations Inclusion in the Canadian Context." *Social Inclusion* 1(1): 21–36.

Wu, Song, Scott Powers, Wei Zhu, and Yusuf A. Hannun. 2016. "Substantial contribution of extrinsic risk factors to cancer development." *Nature* 529: 43–47 (January 7). Retrieved August 9, 2016 (http://www.nature.com/nature/journal/v529/n7584/full/nature16166.html).

Yahoo! Answers. "How many players the NHL has currently?" Retrieved August 8, 2016 (https://answers.yahoo.com/question/index?qid=20100104123254AA3jTeI).

Yamani, Mai, ed. 1996. *Feminism and Islam*. New York: New York University Press.

Young, T. Kue. 2015 [2006]. "Health of Indigenous People." *Canadian Encyclopedia*. Retrieved December 30, 2016 (http://www.thecanadianencyclopedia.ca/en/article/aboriginal-people-health).

Zogby, James J. 2002. *What Arabs Think: Values, Beliefs and Concerns.* Utica, NY: Zogby International/The Arab Thought Foundation.

Index